VANISHING BORDERS

Other Norton/Worldwatch Books

Lester R. Brown et al.

State of the World 1984 through *2000* (an annual report on progress toward a sustainable society)

Vital Signs 1992 through *1999* (an annual report on the environmental trends that are shaping our future)

Lester R. Brown
Christopher Flavin
Sandra Postel
Saving the Planet

Alan Thein Durning
How Much Is Enough?

Sandra Postel
Last Oasis

Lester R. Brown
Hal Kane
Full House

Christopher Flavin
Nicholas Lenssen
Power Surge

Lester R. Brown
Who Will Feed China?

Lester R. Brown
Tough Choices

Michael Renner
Fighting for Survival

David Malin Roodman
The Natural Wealth of Nations

Chris Bright
Life Out of Bounds

Lester R. Brown
Gary Gardner
Brian Halweil
Beyond Malthus

Sandra Postel
Pillar of Sand

VANISHING BORDERS

Protecting the Planet in the
Age of Globalization

HILARY FRENCH

W·W· NORTON & COMPANY
New York   London

363.7
F874v

/01

CONTENTS

ACKNOWLEDGMENTS

In 1972, Lester Brown published a book that was ahead of its time: *World Without Borders*. Nearly 30 years later, the message of that volume still rings true. I am indebted to Lester for his vision of a humane and environmentally sustainable world—a vision that he expressed in that early book and has continued to put forth as Founder and President of Worldwatch Institute. More specifically, I am grateful to Lester for his enthusiasm for this project and, over the years, for the broader program of research from which it sprung.

Undertaking a book of the breadth of *Vanishing Borders* is a daunting task. It could not have been accomplished without extraordinary support from my colleagues at the Worldwatch Institute. This book cuts across the research programs of the Institute, and I have been fortunate to be able to draw on this rich storehouse of knowledge.

I would particularly like to thank Staff Researcher Lisa Mastny for providing key assistance with all aspects of this book. Lisa helped collect, organize, and analyze an often overwhelming amount of information. Her research talents, and her good cheer under fire, were much appreciated. I am also deeply grateful to freelance editor Linda Starke, both for her skillful editing and for her patience and flexibility in working with an author with a penchant for pushing deadlines to their limits. My thanks to Art Director Elizabeth Doherty for producing a beautiful design for the book and for helping make our tight schedule work; to Molly O'Meara and Payal Sampat for pitching in with critical last-minute proofreading assistance; to Librarian Lori Brown for help tracking down obscure documents; and to Suzanne Clift for offering both administrative and moral support.

Most of my colleagues on the Worldwatch staff read various pieces of this book in different forms at different times, including as a draft *State of the World* chapter. I am grateful to all of them. For their comments, I thank Janet Abramovitz, Dick Bell, Chris Bright, Lester Brown, Seth Dunn, Christopher Flavin, Gary Gardner, Brian Halweil, Ashley Mattoon, Anne Platt McGinn, Michael Renner, David Malin Roodman, and Curtis Runyan. Outside the Institute, I am indebted to reviewers Ricardo Bayon, Michelle Chan-Fishel, David Downes, and Frances Seymour for offering useful input on short notice. This book is much the better for the collective contributions of this group.

Now that *Vanishing Borders* is nearing completion, I am finally able to contemplate the outreach process. I look forward to working with a talented Worldwatch team—Dick Bell, Mary Caron, Christine Stearn, and Liz Hopper—in helping this book find its audience, and with Reah Janise Kauffman in offering it to publishers outside the United States. And at W.W. Norton & Company, our U.S. publisher, I appreciate the dependable talents of Amy Cherry, Andrew Marasia, and Nomi Victor.

As always, all of us at Worldwatch are grateful to the various foundations that support the Institute's work: the

Geraldine R. Dodge, Ford, William and Flora Hewlett, W. Alton Jones, the John D. and Catherine T. MacArthur, Charles Stewart Mott, Curtis and Edith Munson, David and Lucile Packard, Summit, Turner, Wallace Genetic, Weeden, Wege, and Winslow Foundations; Rockefeller Financial Services; the U.N. Population Fund; and the Wallace Global Fund. We are also indebted to the Institute's Council of Sponsors—Tom and Cathy Crain, Vicki and Roger Sant, Robert Wallace, and Eckart Wintzen—who contribute generously each year to the Institute.

Finally, I am especially grateful to three individuals. My parents, Alan and Mary French, helped inspire this book in innumerable ways, including childhood hiking trips, visits to the United Nations, and stimulating dinner table conversations. And Christopher Flavin not only provided key substantive input but also, and even more important, offered unwavering encouragement and support. My heartfelt thanks to all.

Hilary French

Worldwatch Institute
1776 Massachusetts Ave., N.W.
Washington, D.C. 20036

December 1999

VANISHING BORDERS

CHAPTER 1

ONE WORLD?

In late November 1999, trade ministers from 135 countries assembled in Seattle to launch a new round of global trade talks. But things did not go according to plan. Instead, delegates to the World Trade Organization (WTO) meeting were greeted by tens of thousands of demonstrators from around the world who delayed the start of the talks through a massive street protest that kept delegates from the convention hall. Sadly, the event turned violent when a handful of renegades used the occasion to launch a spree of random violence, and police sprayed tear gas and fired rubber bullets at the protesters. By the end of the week, hundreds of demonstrators were in jail, mainly for the relatively innocuous offense of blocking public streets. But the official meeting was also in tatters, with delegates scurrying for airplanes without having agreed even to a pro forma official declaration.[1]

The "battle of Seattle," as it was quickly dubbed, may have marked a critical turning point. "If there is any clear message coming through the clouds of tear gas and broken glass in Seattle this week, it is that the terms of the debate about free trade have changed," reported the *Washington Post*. "It is no longer a debate about trade at all, but rather a debate about globalization, a process that many now understand affects not only traditional economic factors such as jobs and incomes but also the food people eat, the air they breathe...and the social and cultural milieu in which they live." Concern about the environmental implications of the WTO and broader globalization trends were high on the list of the concerns of the protesters.[2]

As the controversy swirling around the Seattle meeting made clear, "globalization" has become a contentious process. Part of the conflict stems from the fact that the term means vastly different things to different people. To some, globalization is synonymous with the growth of global corporations whose far-flung operations transcend national borders and allegiances. To others, it signals a broader cultural and social integration, spurred by mass communications and the Internet. The term can also refer to the growing permeability of international borders to pollution, microbes, refugees, and other forces.[3]

Globalization is used here to refer to a broad process of societal transformation that encompasses all of the above, including growth in trade, investment, travel, computer networking, and transboundary pollution. (See Table 1–1.) This book explores the collective impact of these phenomena on the health of the planet's natural systems.[4]

Today's integrated world is the result of a process that can be traced back 1 million years, when early humans first migrated out of Africa throughout Eurasia. It was not until the 1500s, however, that people living several continents apart came into contact as a result of the European Age of Exploration. The late nineteenth century brought the development of steam-powered ships and railroads, which dramatically expanded international commerce and exchange. Two World Wars and the Great Depression slowed globalization dramatically in the first half of the twentieth century. But the second half brought globalization back with abandon, as trade rebounded and widespread international air travel and the use of personal computers revolutionized links between countries and cultures.[5]

Growth in trade has consistently outpaced the expansion of the global economy since World War II. The world economy has grown sixfold since 1950, rising from $6.7 trillion to $41.6 trillion in 1998. But exports increased 17-fold over this period, reaching $5.4 trillion in 1998. (See Figure 1–1.) While exports of goods accounted for only 5 percent of the gross world product in 1950, by 1998 this figure had climbed to 13 percent.[6]

In recent decades, international investment by multinational corporations has also exploded. Over the 1980s, foreign direct investment flows grew twice as fast as trade—increasing 15-fold between 1970 and 1998, from $44 billion to $644 billion. The number of transnational corporations (TNCs) has also soared in recent decades, increasing from only 7,000 in 1970 to more than 53,000 in 1998. And not only companies are now investing abroad. Some 44 million U.S. households have at least some money in mutual funds, up from only 4.6 million in 1980. Their dollars are increasingly invested overseas: the assets of U.S.-based international and global mutual funds climbed from just $16 billion in 1986 to $321 billion at the end of 1996.[7]

The globalization of commerce in recent decades has internationalized environmental issues. Trade in natural resources such as timber and fish is soaring. Common trappings of daily life—a teak coffee table, for instance, or a salmon dinner—can affect the well-being of people and ecosystems on the other side of the world. And international investments are giving millions of people an influence, albeit often unwitting, on environmental developments in distant corners of the planet.[8]

TABLE 1-1

Globalization at a Glance

Indicator	Trend
World Trade	Between 1950 and 1998, world exports of goods increased 17-fold—from $311 billion to $5.4 trillion—while the global economy expanded only sixfold. Exports of services have also surged in recent decades—from $467 billion in 1980 to $1.3 trillion in 1997—and now represent nearly one fifth of total world trade.
Private Investment/ Capital Flows	Between 1970 and 1998, global foreign direct investment increased from $44 billion to $644 billion. Capital flows to developing countries alone grew 11-fold between 1970 and 1998, from $21 billion to $227 billion. The share of capital entering the developing world from private sources doubled over this period, reaching 88 percent.
Transnational Corporations (TNCs)	Between 1970 and 1998, the number of TNCs worldwide grew from 7,000 to an estimated 53,600, with some 449,000 foreign subsidiaries. The sales of TNCs outside their home countries are growing 20–30 percent faster than their exports, and sales of goods and services by foreign subsidiaries—valued at $9.5 trillion in 1997—surpass total world exports by nearly 50 percent.
Shipping	Between 1955 and 1998, the tonnage of goods carried by ship rose more than sixfold, to 5.1 billion. Meanwhile, the unit cost of carrying freight by ship dropped 70 per-cent between 1920 and 1990 (in 1990 dollars).
Air Transport	Between 1950 and 1998, the number of passenger-kilo-meters flown internationally grew nearly 100-fold, from 28 billion to 2.6 trillion. Air freight also soared over this period, from 730 million to 99 billion ton-kilometers carried. Mean-while, the average revenue per mile for air transport fell from 68¢ to 11¢ between 1930 and 1990 (in 1990 dollars).
Tourism	Between 1950 and 1998, international tourist arrivals increased 25-fold, from 25 million to 635 million. Some 2 million people now cross an international border each day, compared with only 69,000 in 1950.
Refugees	Between 1961 and 1998, the number of international refugees qualifying for and receiving U.N. assistance grew 16-fold, from 1.4 million to 22.4 million. Today, the

TABLE 1–1 *(continued)*

	total number of refugees worldwide—including internally displaced persons, asylum seekers, and people living in refugee-like situations—tops 56 million.
Telephones	Between 1960 and 1998, the number of lines linking non-cellular telephones directly to the global phone network grew eightfold, from 89 million to 838 million. In developing countries, the number of phone connections per 100 people jumped from only 1 in 1975 and 2 in 1985 to 6 in 1998. Meanwhile, the average cost of a three-minute phone call from New York to London fell from $244.65 in 1930 to $3.32 in 1990 (in 1990 dollars).
Internet/ Computing	Since 1995, the Internet has grown by roughly 50 percent each year, following 15 years of more than doubling in size annually. In 1998, some 43 million host computers wired an estimated 147 million people to the Internet. Today, 1 in every 40 people has access. Meanwhile, the unit cost of computing power fell 99 percent between 1960 and 1990 (in 1990 dollars).
Nongovern-mental Organizations (NGOs)	Between 1956 and 1998, the number of international NGOs (groups operating in at least three countries) grew 23-fold, from only 985 to an estimated 23,000. A study of 22 nations worldwide found that the nonprofit sector accounted for 5.7 percent of the national economy on average and employed 5 percent of the total workforce.

SOURCE: See endnote 4.

A biotic intermingling of unprecedented proportions is also taking place as species and microbes that were once neatly contained within geographic boundaries are now let loose by trade and travel. And wind and ocean currents, rainfall, rivers, and streams carry contaminants hundreds or even thousands of miles from their sources. DDT and PCBs, for instance, have been found throughout the Inuit food chain in the Arctic, from the snow and edible berries to fish and polar bears. On an even larger scale, ozone depletion, climate change, and oceanic pollution threaten all nations.[9]

The unparalleled economic expansion after World War II

FIGURE 1–1

World Export of Goods, 1950–98

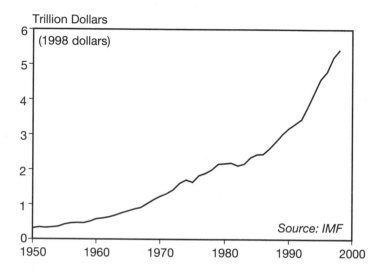

brought with it a burst in the consumption of material goods. Global wood consumption has more than doubled since 1950, paper use has increased sixfold, fish consumption has grown fivefold, water and grain consumption have tripled, and steel use and fossil fuel burning have climbed fourfold. The world has also seen an explosion in human numbers. The number of people inhabiting the planet has more than doubled—from 2.5 billion in 1950 to 6.0 billion in 1999.[10]

The combination of these trends has caused the world economy to begin to push up against the planet's ecological limits. In 1998, the carbon emissions that are one of the main causes of global warming were near their peak (see Figure 1–2), and carbon dioxide concentration in the atmosphere again reached record levels. Biologists warn that we have entered a period of mass extinction of species—the largest die-off in 65 million years. According to surveys by

FIGURE 1–2

World Carbon Emissions From Fossil Fuel Burning, 1950–98

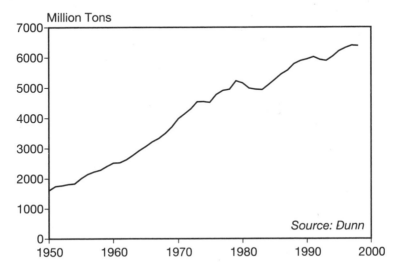

the World Conservation Union–IUCN, an estimated one quarter of the world's mammal species are threatened with extinction, as are nearly 13 percent of plant species. The world's major fisheries are on the verge of collapse, and water scarcity and land degradation threaten our ability to feed the more than 6 billion people that now inhabit the planet.[11]

The global nature of both the economy and of ecological systems causes the exchange of "environmental space" among nations. A team of researchers led by Mathis Wackernagel of the Center for Sustainability Studies in Xalapa, Mexico, has calculated what they call the "ecological footprint" of 52 nations: the amount of biologically productive land area appropriated by these countries and their inhabitants. When all 52 are tallied up, it becomes clear that the world is already living beyond its ecological means. But some countries are doing so far more than others as a result of either scarce natural capital, profligate consumption patterns, or some combination of the two. (See Table 1–2.) Countries in ecological deficit import natural capital from those in surplus, an element of globalization that few people are conscious of.[12]

As environmental concerns become more pressing, they are climbing higher on the international political agenda. The Seattle meeting demonstrated that global economic negotiations that ignore ecological issues do so at their peril. But global eco-politics is becoming increasingly strained. Industrial countries often disagree among themselves, with the European Union and the United States now at odds on issues ranging from global climate change to genetically modified organisms. Environmental issues have also become acrimonious in North-South relations, with rich and poor countries divided over how to address these issues in the context of the global economy, and over how to apportion responsibility for reversing the planet's ecological decline.

Globalization in its many guises poses enormous challenges to traditional governance structures. National governments are ill suited for managing environmental problems that transcend borders, whether via air and water currents or through global commerce. Yet international environmental governance is still in its infancy, with the treaties and institutions that governments turn to for global management mostly too weak to put a meaningful dent in the problems. Nations are granting significant and growing powers to economic institutions such as the WTO and the International Monetary Fund, but environmental issues remain mostly an afterthought in these bodies, despite the best efforts of demonstrators and public policy groups.

While nation-states are losing ground in the face of globalization, other actors are moving to the fore, particularly international corporations and nongovernmental organizations. New information and communications technologies

TABLE 1–2

Ecological Footprint Per Person in Selected Nations, 1995

Country	Available Ecological Capacity	Ecological Footprint	Ecological Deficit or Surplus (Capacity minus Footprint)
	(hectares per capita)		
Netherlands	1.2	5.9	– 4.7
United States	6.7	10.9	– 4.2
Japan	0.8	4.7	– 3.9
Israel	0.3	3.7	– 3.5
South Korea	0.4	3.8	– 3.4
United Kingdom	1.8	4.9	– 3.1
Greece	1.8	4.8	– 3.0
Germany	1.9	4.8	– 2.9
South Africa	1.3	3.1	– 1.8
France	4.0	5.4	– 1.4
Mexico	1.4	2.6	– 1.2
China	0.6	1.5	– 0.8
India	0.5	1.0	– 0.5
Russia	4.3	4.7	– 0.5
Indonesia	2.7	1.4	1.3
Canada	12.6	7.4	5.2
Brazil	9.1	3.8	5.3
Australia	16.3	10.0	6.3
Iceland	21.8	6.6	15.2
New Zealand	26.8	8.2	18.6
World	2.0	2.4	– 0.4

SOURCE: Mathis Wackernagel and Alejandro Callejas, "The Ecological Footprints of 52 Nations (1995 data)," Redefining Progress, available at <www.rprogress.org>.

are facilitating international networking, and activist groups, businesses, and international institutions are forging innovative partnerships.

But though the economy and the environment are both increasingly global, politics continues to be mostly national and local. As Professor Dani Rodrik of Harvard University puts it: "Markets are sustainable only insofar as they are embedded in social and political institutions....It is trite but true to say that none of these institutions exists at the global level."[13]

The world economy and the natural world that it relies on are both in precarious states as we enter the new millennium, provoking fears that an era of global instability looms on the horizon. Over the course of the twentieth century, the global economy stretched the planet to its limits. The time is now ripe to build the international governance structures needed to ensure that the world economy of the twenty-first century meets peoples' aspirations for a better future without destroying the natural fabric that underpins life itself.

I

THE ECOLOGY OF GLOBALIZATION

CHAPTER 2

NATURE UNDER SIEGE

While economists tout record-breaking increases in global commerce in recent decades, more sobering statistics are being put forth by the world's leading biologists: the loss of living species in recent decades, they report, represents the largest mass extinction since the dinosaurs were wiped out 65 million years ago.[1]

Globalization is a powerful driving force behind today's unprecedented biological implosion. Trade in timber, minerals, and other natural commodities is climbing, and many of the world's hotspots of biological diversity are now threatened by a surge of international investment in resource extraction. (See Figure 2–1.) Yet the new rules of the global economy pay little heed to the importance of reversing the biological impoverishment of the planet. This mismatch between ecological imperatives and prevailing economic practice will need to be bridged if the world is to avoid an unraveling of critical environmental services in the early part of this new century.[2]

Human beings remain fundamentally dependent on the

FIGURE 2–1

Resource Investments and Natural Areas

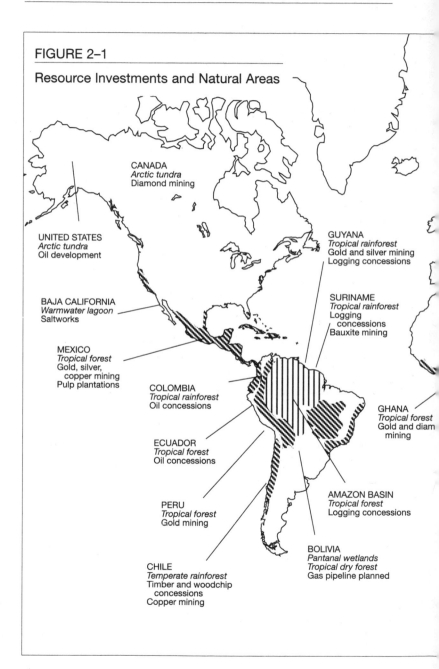

CANADA
Arctic tundra
Diamond mining

UNITED STATES
Arctic tundra
Oil development

GUYANA
Tropical rainforest
Gold and silver mining
Logging concessions

BAJA CALIFORNIA
Warmwater lagoon
Saltworks

SURINAME
Tropical rainforest
Logging
   concessions
Bauxite mining

MEXICO
Tropical forest
Gold, silver,
   copper mining
Pulp plantations

COLOMBIA
Tropical rainforest
Oil concessions

GHANA
Tropical forest
Gold and diam
   mining

ECUADOR
Tropical forest
Oil concessions

PERU
Tropical forest
Gold mining

AMAZON BASIN
Tropical forest
Logging concessions

BOLIVIA
Pantanal wetlands
Tropical dry forest
Gas pipeline planned

CHILE
Temperate rainforest
Timber and woodchip
   concessions
Copper mining

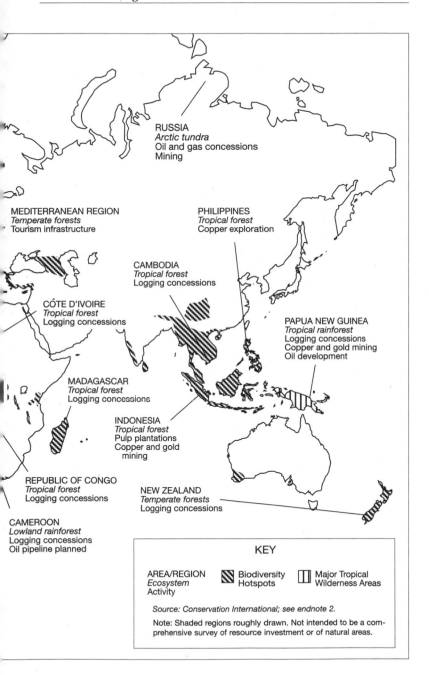

RUSSIA
Arctic tundra
Oil and gas concessions
Mining

MEDITERRANEAN REGION
Temperate forests
Tourism infrastructure

PHILIPPINES
Tropical forest
Copper exploration

CAMBODIA
Tropical forest
Logging concessions

CÔTE D'IVOIRE
Tropical forest
Logging concessions

PAPUA NEW GUINEA
Tropical rainforest
Logging concessions
Copper and gold mining
Oil development

MADAGASCAR
Tropical forest
Logging concessions

INDONESIA
Tropical forest
Pulp plantations
Copper and gold
   mining

REPUBLIC OF CONGO
Tropical forest
Logging concessions

NEW ZEALAND
Temperate forests
Logging concessions

CAMEROON
Lowland rainforest
Logging concessions
Oil pipeline planned

KEY

AREA/REGION
Ecosystem
Activity

Biodiversity
Hotspots

Major Tropical
Wilderness Areas

Source: Conservation International; see endnote 2.

Note: Shaded regions roughly drawn. Not intended to be a comprehensive survey of resource investment or of natural areas.

natural world. One shortcoming of conventional economics is its failure to account for the critical services provided by natural ecosystems such as forests, wetlands, coral reefs, rivers, and seas. In 1997, a team of 13 ecologists, economists, and geographers published a path-breaking article that put a price tag on the value of a range of functions provided by these ecosystems. The study covered a broad array of services, including genetic resources, flood control, pollination, water supply, and erosion control. The authors arrived at the stunning conclusion that the economic value of "nature's services" adds up to some $33 trillion each year—almost as much as the entire annual gross world product.[3]

Despite their value to humankind, ecosystems are being degraded at an unparalleled rate as a result of human activity. One benchmark of the losses is the rapid rate at which species are being extinguished. Biologists warn that as many as one fifth of all plant and animal species could disappear within the next 30 years. Another measure of ecological health is the extent to which humans have transformed ecosystems from their natural state into cropland, pasture, plantations, human settlements, and other uses. Many countries have seen already seen more than half of their land area undergo this conversion, including Argentina, Australia, India, Mexico, South Africa, and Spain.[4]

Nations ostensibly set about the task of staunching biological losses at the U.N. Conference on Environment and Development in Rio in 1992, when they finalized a U.N. Convention on Biological Diversity. The accord has now been ratified by more than 175 countries (although the United States is not one of them). Among its many provisions, the treaty requires countries to adopt national biodiversity strategies and action plans, establish protected areas, conserve threatened species, restore degraded habitats, and fairly and equitably share the benefits of genetic resources. But unlike the rules of the World Trade Organization (WTO), the biological diversity convention contains few concrete commitments and no effective enforcement mechanisms. Not surprisingly, it has so far failed to put a measurable dent in the burgeoning global extinction crisis. Reversing ecological decline will require going beyond exhortation to weave biological integrity into the fabric of the global economy.[5]

THE TIMBER TRADE

The world's forests are a particularly important reservoir of biological wealth. They harbor more than half of all species on Earth and provide a range of other important natural services, including flood control and climate regulation. But the planet's forest cover is steadily shrinking as human numbers and the global economy continue to expand. Nearly half of the forests that once covered Earth have already been lost, and almost 14 million hectares of tropical forest—an area nearly three times the size of Costa Rica—is being destroyed each year.[6]

The role of international trade in global deforestation has been a subject of controversy over the years. Global timber trade is far from the only culprit in forest loss: the clearing of land for agriculture and grazing is also a major cause, as are fuelwood gathering and the felling of trees for domestic use. Yet the draw of international markets can be an inducement for countries to cut down trees far faster than would be required to meet domestic demand alone. Several countries export more wood than they consume domestically, including Cameroon, Canada, Gabon, and Papua New Guinea. Indonesia and Malaysia have both pushed plywood exports with gusto in recent years, contributing in no small measure to rapid deforestation in both countries. Plywood exports from the two countries combined exploded from just 233,000 cubic meters in 1975 to 12 million cubic meters in 1998. These two countries now account for nearly 60 percent of world plywood exports, up from just 4 percent in 1975.[7]

Recent studies have concluded that commercial logging is the preeminent threat to the world's most biologically rich forests. High on this list are the "frontier forests" of Alaska, Canada, Russia, the Amazon Basin, and the Guyana Shield— the world's last remaining tracts of relatively undisturbed natural forests. A 1997 World Resources Institute report concluded that nearly 40 percent of the world's frontier forests are threatened by ongoing or planned human activities, with logging implicated in more than 70 percent of the cases. Commercial logging often sets in motion the destruction of a far larger area, as road construction opens the way to mining, farming, hunting, and other disruptive activities. Timber operations also contribute to forest destruction by displacing local peoples, concentrating them on smaller forest fragments that are less able to supply fuelwood and fodder at sustainable rates.[8]

The value of global trade in forest products has risen steadily over the last few decades, climbing from $29 billion in 1961 to $139 billion in 1998. (See Figure 2–2.) Recent years have seen particularly rapid growth in trade in more finished types of forest products such as plywood, pulp, and paper. Between 1970 and 1998, exports of wood-based panels increased fivefold in volume, plywood and paperboard exports quadrupled, and exports of sawnwood (a more processed type of wood used in construction) and wood pulp both doubled, according to U.N. Food and Agriculture Organization estimates. Exports of industrial roundwood (raw logs), in contrast, have remained relatively constant. For all products other than logs, exports as a share of total world production increased significantly over this period— an important indication of the growing globalization of the industry. (See Table 2–1.) This growth trend is expected to continue, in response to growing competition, market segmentation, and trade liberalization, among other forces.[9]

Industrial countries are the dominant players in forest

FIGURE 2–2

World Trade in Forest Products, 1961–98

products trade, accounting for roughly 80 percent of the value of both exports and imports. (See Table 2–2.) But developing countries have steadily increased their share of plywood, pulp and paperboard, and other forest products exports over the last few decades. Because most net deforestation today is taking place in a handful of biologically rich tropical countries, forest products exports from these nations have particular significance for global biodiversity loss. Brazil, Indonesia, and Malaysia have all now joined the ranks of the top 10 forest products exporters. But their forests have paid a heavy price for this export success— these three countries alone accounted for some 40 percent of global forest loss during the 1980s and 36 percent of the loss in the first half of the 1990s.[10]

Officially reported trade represents just the tip of the iceberg, as much of the international timber trade is illegal, conducted in the shadows. Bolivia, Brazil, Cambodia,

TABLE 2–1

Trade in Selected Wood and Wood Products, 1970 and 1998

Product	1970		1998	
	Exports[1] (million)	Exports as Share of Production (percent)	Exports[1] (million)	Exports as Share of Production (percent)
Wood Pulp	17	17	114	27
Paper and Paperboard	23	18	90	31
Industrial Roundwood	94	7	85	6
Plywood	5	15	53	34
Sawnwood	57	14	34	21
Wood-Based Panels	10	14	21	40

[1]All units in cubic meters except wood pulp and paper and paperboard, which are in tons.
SOURCE: Worldwatch Institute, based on data in U.N. Food and Agriculture Organization, *FAOSTAT Statistical Database*, electronic database, <apps. fao.org>, viewed 22 October 1999.

Cameroon, Ecuador, Georgia, Ghana, Indonesia, Kenya, Laos, Mexico, Paraguay, the Russian Federation, Thailand, and Viet Nam are among the countries where illegal trade in timber plays an important role in the decimation of forests.[11]

For many years now, companies from countries with depleted forests have been turning their chain saws loose overseas. European firms, for instance, have long been active in Africa: in the early 1980s, some 90 percent of logging operations in Gabon were foreign-owned, as were some 77 percent of those in Congo, nearly 90 percent of those in Cameroon, and virtually all of those in Liberia. At least 17 European companies were operating in Côte d'Ivoire alone

TABLE 2–2

Top 10 Exporters and Importers of Forest Products
by Value, 1998

Country	Exports	Share of World Total
	(billion dollars)	(percent)
Canada	25	18
United States	18	13
Finland	11	8
Germany	10	7
Sweden	10	7
Indonesia	5	4
Austria	4	3
Malaysia	4	3
Russian Federation	3	2
Brazil	3	2
World	139	100

Country	Imports	Share of World Total
	(billion dollars)	(percent)
United States	24	16
Japan	17	12
Germany	11	8
United Kingdom	10	7
Italy	9	6
China	9	6
France	7	5
Canada	5	3
Netherlands	5	3
South Korea	4	3
Hong Kong	4	3
World	148	100

SOURCE: Worldwatch Institute, based on data in FAO, *FAOSTAT Statistical Database*, electronic database, <apps.fao.org>, viewed 22 October 1999.

in 1990, and they have shown no sign of letting up in recent years. Japanese firms, for their part, joined forces with local companies in the 1970s and 1980s to decimate the forests of Southeast Asian countries such as Indonesia and Malaysia.[12]

With their own forests greatly reduced, logging companies from Indonesia, Malaysia, and elsewhere in Asia have themselves begun investing abroad. Asian companies have in recent years purchased vast timber concessions in Africa, Asia, and North and South America that threaten some of the world's last remaining untouched forests. Brazil, Cameroon, the Democratic Republic of Congo, Guyana, Papua New Guinea, and Suriname are among the countries that have sold foreign investors the rights to log large tracts of primary forests—often at prices that do not reflect the marketplace value let alone the ecological worth of these areas. Some of the companies involved have a long record of catastrophic environmental destruction as well as corruption—a fact that does not bode well for the countries that extended the welcome mat.[13]

International companies are also stepping up their investments in related wood-products industries such as sawmills and pulp and paper operations that feed off of steady streams of locally supplied wood. Some 15 U.S. wood-products companies have set up shop in Mexico since the North American Free Trade Agreement was ratified in 1994. And in Argentina, Brazil, Chile, China, and Indonesia, multinational companies have joined forces with local investors to produce wood chips and pulp and paper at mills supplied by vast monoculture tree plantations that are being planted at a rapid rate. Japanese companies are major players in this business, with pulp and paper operations located in Australia, Canada, Chile, China, and Papua New Guinea, among other places.[14]

A controversial proposed World Trade Organization agreement on liberalizing trade in forest products could add to the pressures that global commerce is placing on the world's forests. Under the agreement now being considered, most industrial countries would eliminate tariffs on pulp and paper by 2000, and on wood and other forest products such as furniture by 2002. Developing countries would be given an additional two years to meet these terms. The precise effects of these steps are difficult to predict, but studies suggest that the higher prices paid to producers as a result of tariff reductions will boost production in some countries. A recent U.S. government report concluded that the agreement would likely increase production by nearly 3 percent in Malaysia and over 4 percent in Indonesia, although the report also forecasts production declines in some countries, including Mexico and Russia. With so little of today's timber industry based on sustainable practices, production increases often translate into increased forest destruction.[15]

Although the proposed accord would initially take aim only at tariffs, its scope might well be expanded later to include so-called nontariff barriers to trade. Over the longer term, these provisions might pose an even greater threat to the health of the world's forests, and to the diversity of species that inhabit them. For instance, forest certification initiatives aimed at creating a market for sustainably harvested timber could run head-on into WTO rules in the years ahead. A recent report by Asia-Pacific Economic Cooperation, a regional trade grouping, flagged a number of important forest protection policies as potential nontariff trade barriers, including a ban on logging in China's upper Yangtze basin that was instituted in response to recent catastrophic flooding in the region.[16]

MINING THE EARTH

Mining and petroleum development also threaten the health of the world's forests, mountains, waters, and other sensitive ecosystems. Mining exacts enormous environmental costs,

ranging from the destruction of huge tracts of land to the
generation of prodigious quantities of pollution and waste.
For every kilogram of gold produced in the United States,
for example, some 3 million kilograms of waste rock are left
behind. Prime extraction sites are often located in previously undisturbed forests or wilderness areas. According to the
World Resources Institute, mining, energy development,
and associated activities represent the second biggest threat
to frontier forests after logging, affecting nearly 40 percent of
threatened forests.[17]

Besides disturbing valuable ecosystems, mining also can
be devastating for local people: by one estimate, 50 percent
of the gold produced in the next 20 years will come from
indigenous peoples' lands. Toxic byproducts of mining poison the rivers that local people drink from, and the mining
operations themselves destroy the forests and fields that provide sustenance.[18]

Industrial countries are the main consumers of minerals,
accounting for nearly 100 percent of nickel imports, more
than 90 percent of bauxite imports, over 80 percent of zinc
imports, and roughly 70 percent of copper, iron, lead, and
manganese imports. But it is developing countries that are
the main exporters of mineral resources, and that are most
at risk from the associated environmental damage. Collectively, developing countries account for 76 percent of all
exports of bauxite and nickel ore, 67 percent of copper, 54
percent of tin, and 45 percent of iron ore.[19]

In recent years, minerals exploration has slowed in traditional mining countries while picking up in many parts of
the developing world. From 1991 to 1999, spending on
exploration for nonferrous metals more than tripled in Latin
America and grew slightly in Africa and in the Pacific region,
while declining steeply in North America. Nearly 30 percent
of spending on mineral exploration currently takes place in
Latin America, now the leading region, up from just 11 per-

cent in 1991. (See Table 2–3.)

The U.S. mining industry blames environmentalists for the migration, arguing that tighter environmental regulations have made domestic mining a difficult and expensive proposition. More significant is the fact that host countries are inviting international investors in with open arms; some 70 countries have rewritten their national mining codes in recent years with the aim of encouraging investment. Yet few are devoting similar energy to strengthening environmental laws and enforcement.[20]

Like mining companies, multinational oil and gas firms continually scour the planet for new development opportunities, as the most accessible fields in industrial countries have already been tapped. More than 90 percent of known

TABLE 2–3

Worldwide Metals Exploration Spending, by Location, 1991 and 1999[1]

Region	1991 Amount (million dollars)	1991 Share (percent)	1999 Amount (million dollars)	1999 Share (percent)
Latin America	200	11	630	29
North America	771	41	450	21
Australia	353	19	404	19
Africa	315	17	323	15
Pacific Region	125	7	175	8
Rest of World[2]	82	4	182	8
Total[3]	1,846	100	2,170	100

[1]Includes precious, base, and other non-ferrous hard-rock metals; based on the budgets of major mining companies that represent 81 percent of worldwide metals exploration spending. [2]Includes Europe, former Soviet Union, Middle East, and Asia (excluding Pacific nations). [3]Share columns may not add up to 100 percent due to rounding.
SOURCE: Metals Economic Group (MEG), *Strategic Report* (Halifax, NS, Canada: November/December 1991); MEG, "A 23% Decrease in 1999 Exploration Budgets," press release (Halifax, NS, Canada: 20 October 1999).

oil reserves and nearly 60 percent of natural gas reserves are
located in the developing world. The Middle East is still the
dominant region for oil and gas, but the major companies
are also increasingly striking deals—and oil—in the Central
Asian republics, deep in the South American rainforest, and
off Asian and West African shores. As with mineral extraction, the environmental and social costs are high when previously remote and pristine areas are opened up to
development.[21]

HARNESSING NATURAL WEALTH

There can be little doubt that globalization has accelerated
the unprecedented loss of biological riches in recent
decades. But this tragic connection is not immutable. A variety of alternative policies and practices now being tried
around the world could be scaled up to create a global economy that nurtures rather than decimates natural wealth.

One promising approach is to harness consumer power
on behalf of environmental change. The last few years have
seen a flurry of activity aimed at encouraging more sustainable timber harvesting through certification and eco-labeling
programs. The pioneer in these efforts is the Forest Stewardship Council (FSC), an independent body established in
1993 to set standards for sustainable forest production
through a cooperative process involving timber traders and
retailers as well as environmental organizations and forest
dwellers. Although certified timber currently accounts for
only a small share of all timber production, demand for this
product is growing fast, as is the number of certified forest-
lands. As of late 1999, FSC-accredited bodies had certified
some 17 million hectares of forest, up from only 1 million
hectares in late 1995. More than 70 percent of all certified
forests are located in just three countries—Sweden, Poland,
and the United States. But certified forests can be found in

30 countries overall, including Brazil, Canada, Malaysia, and the Solomon Islands. In 1998, the World Bank entered into a partnership with the World Wildlife Fund aimed at boosting the number of hectares of independently certified forests to 200 million worldwide (100 million each in the tropical and temperate regions) by 2005.[22]

Another strategy for preserving forests while providing a livelihood to those who live in them is to promote trade in nontimber forest products such as nuts, rattan (palm stems used for wicker furniture and baskets), rubber, and spices. International trade in these products is already substantial— some $11 billion annually, according to the U.N. Food and Agriculture Organization. Although there is no guarantee that alternative forest products will be harvested sustainably, they are more likely to be than timber. This approach was first pursued by Brazil's rubber tappers, who have worked to attain rights to "extractive reserves" where these nontimber forest products can be produced on a sustainable basis and sold to companies like Ben and Jerry's and The Body Shop, which derive environmental cachet for their products by their association with this cause.[23]

Carefully controlled tourism is another possible means for channeling funds into the preservation of threatened ecosystems, although without care it can also be a quick route toward their destruction. International tourism has climbed rapidly in recent decades, as air travel has become steadily cheaper, disposable income and leisure time have grown, and tourist destinations have generally become more accessible. In 1950 there were 25 million international tourist arrivals worldwide. By 1998 this number had grown 25-fold, to 635 million. The World Tourism Organization projects that international arrivals will reach 1.6 billion by 2020, an increase of 250 percent over 1998. "Nature tourism" is one of the fastest-growing segments of the industry—accounting for some 40–60 percent of the total, depending on how it is defined. The tourism group projects that the trendiest vacation destinations in the new millennium will be "the tops of the highest mountains, the depths of the oceans, and the ends of the Earth."[24]

The growing reach of international tourism threatens to put great strains on sensitive environments, and to contribute to the erosion of threatened cultures. The number of countries offering whale watching excursions, for example, has grown from 10 to 65 in the past two decades, and more than 5 million tourists participate in the activity each year. But these excursions, which often advertise whale "petting" as well as viewing, have contributed to the disturbance and harassment of whale populations in their traditional breeding sites, including the warm waters off Baja California, Argentina, and the Canary Islands. Similarly, ecotourism activity in remote regions of Venezuela has led to conflicts between tour operators and indigenous communities, who claim that tourist camps are pushing them from their customary lands. And all too often the economic benefits accrue to international investors and national treasuries rather than to local people.[25]

Yet a number of countries have succeeded in harnessing revenues from tourism in a way that allows wildlife and wild places to pay their own way, bringing in much needed income for impoverished people. Costa Rica is a case in point. The country's moist cloud forests, sandy beaches, and dry deciduous forests have made tourism the top foreign exchange earner, surpassing traditional export mainstays such as bananas and coffee. The Costa Rican government is working to ensure that this tourism is carried out in an environmentally friendly and culturally sustainable manner through its Certification for Sustainable Tourism program, which categorizes and rates hotels and tourism providers according to how responsibly they operate. And several African nations, including Zimbabwe and Rwanda, have successfully funneled wildlife tourism income to local people, thereby giving them a stake in wildlife preservation.[26]

"Bioprospecting" is another possible strategy for making the preservation of biological diversity pay. Drug and seed companies have long used the genetic diversity of the developing world to create new products. Yet even when a traditional crop variety proves essential for breeding a new line of seeds, or when a wild plant yields a valuable new drug, corporations have rarely paid anything for access to the resource. The Convention on Biological Diversity signed at the Earth Summit in 1992 gives nations the right to charge for access to genetic resources, and it allows them to pass national legislation setting the terms of any bioprospecting agreements. One of the goals of these provisions was to make it more profitable for biologically rich countries to preserve their natural wealth than to destroy it. Looming in the background were rapid advances in biotechnology, which relies on a rich natural storehouse of genetic material and could thus boost bioprospecting returns substantially.[27]

A year before the Earth Summit, Merck and Company, one of the world's largest pharmaceuticals firms, and Costa Rica's Instituto Nacional de Biodiversidad (INBio) reached a precedent-setting bioprospecting agreement roughly along the lines of the arrangement subsequently envisioned in the biodiversity treaty. Merck agreed to pay INBio $1.35 million in support of conservation programs in exchange for access to the country's plants, microbes, and insects, as well as royalty payments when any discovery makes its way into a product. Though widely hailed as an important step forward, the agreement has also generated substantial controversy. Its critics question whether or not the royalty rate was set at a fair level, and the extent to which the economic benefits will reach the local peoples whose knowledge of medicinal properties is so central to making the deal work.[28]

It remains to be seen how much of a growth industry bioprospecting will prove to be. Some observers argue that the large number of samples that need to be screened to yield a commercially valuable product makes the cost of bio-prospecting prohibitive for the private sector. It is also possible that scientific advances such as new chemical techniques that enable molecules to be synthesized in laboratories will reduce the pharmaceutical industry's reliance on natural storehouses.[29]

Nonetheless, a number of other bioprospecting programs are now taking shape—some of which offer more equitable models for the distribution of revenues than the INBio venture. A bioprospecting initiative in Suriname, for instance, involves a number of different partners, including indigenous healers, a Surinamese pharmaceutical company, the U.S.-based Bristol-Myers Squibb company, the environmental group Conservation International, and the Missouri Botanical Garden. Royalties from any drugs developed will be channeled into a range of local institutions, including nongovernmental organizations, the national pharmaceutical company, and the forest service. In addition, a Forest Peoples Fund has been established to support small-scale development projects that benefit local indigenous peoples.[30]

Some countries are also trying to put a price tag on another valuable service provided by standing forests—absorbing and storing carbon. A hectare of moist tropical primary forest in Brazil can hold over 300 tons of carbon, while a hectare of mature Douglas fir forest in Canada can retain over 600 tons. Under the Kyoto Protocol to the climate change treaty (see Chapter 6), countries may be permitted to in effect charge for this service by selling carbon permits to companies or countries interested in offsetting their own carbon emissions with forest preservation projects.[31]

A number of experiments with this approach have already been launched. For example, in January 1997 Costa Rica sold some credits for 200,000 tons of carbon at a price of $10 per ton to the Norwegian government, a consortium of Norwegian companies, and a Chicago-based trading company. The proceeds are being channeled into forest regeneration and protection programs. Some analysts estimate that carbon offset projects could eventually generate tens of billions of dollars annually for forest preservation.[32]

But complicated technical issues must still be resolved if trading in carbon sequestration is to prove meaningful and practical as a strategy for combating climate change. Many all-important details of the Kyoto Protocol are still under negotiation. As currently written, the protocol offers credits for carbon sequestered by planting forests, but does not require debiting for carbon released into the atmosphere as a result of commercial timber operations unless the logging results in the permanent conversion of forestland to other uses. The protocol also does not currently distinguish between monoculture tree plantations and biologically rich natural forests. In a worst-case scenario, the protocol could thus provide a perverse incentive for countries to harvest natural forests and replace them with uniform plantations.[33]

Putting a price tag on nature can help create an incentive to preserve it. But the value of intact ecological systems is in the end beyond measure. Creating a global economy that protects rather than destroys natural wealth is both a moral imperative and a practical necessity as we enter the new century.

CHAPTER 3

THE BIOTIC MIXING BOWL

For most of history, natural boundaries such as mountains, deserts, and ocean currents have served to isolate ecosystems and many of the species they contain. But these physical barricades are now becoming permeable as people and organisms spread around the globe, leading to ecological disruptions with damaging and unpredictable consequences.

Ecological integration has accelerated dramatically in recent decades, as trade and travel have skyrocketed. More than 5 billion tons of goods were shipped across the world's oceans and other waterways in 1998, more than six times as much as in 1955. (See Figure 3–1.) International air travel is also soaring. More people are flying greater distances than ever before, with 2 million people now crossing an international border every day. Since 1950, the number of passenger-kilometers flown has increased at an average annual rate of 9 percent, reaching over 2.6 trillion in 1998. (See Figure 3–2.)[1]

The rapid growth in the movement of human beings and their goods and services around the world has provided convenient transportation for thousands of other species of

FIGURE 3–1

International Seaborne Trade, 1955–98

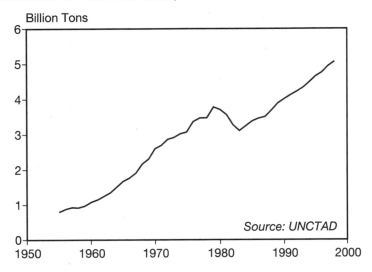

plants and animals that are now taking root on foreign shores. This explosion in the movement of species and microbes across international borders poses a major threat to both the planet's biological diversity and the health of its human inhabitants.[2]

THE BIOINVASION THREAT

The world community is just beginning to awaken to the pervasive danger posed by the spread of non-native "exotic" species, a process dubbed bioinvasion. Once exotics establish a beachhead in a given ecosystem, they often proliferate, suppressing native species. Invasive species are a major threat to the diversity of life on Earth. Nearly 20 percent of the world's endangered vertebrate species are threatened by exotics, and almost half of all species in danger of extinction in the United States are imperiled at least in part by non-native species.[3]

FIGURE 3–2

World Air Travel, 1950–98

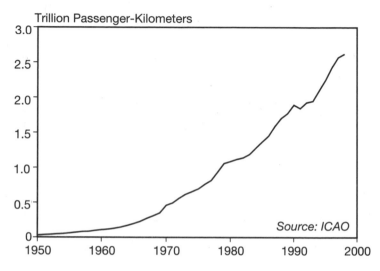

Trillion Passenger-Kilometers

Source: ICAO

Ballast water from international shipping is a major culprit in the spread of aquatic species. On any given day, some 3,000–10,000 aquatic species are moving around the world in ship ballasts. When the ballast water is discharged, so are the organisms, after which they often cause incalculable damage. For example, a ballast water–induced invasion of the Black Sea by the Atlantic jellyfish in the early 1980s was instrumental in the collapse of the fisheries there by the end of that decade.[4]

The U.S. Great Lakes have also been hard hit by bioinvasions over the last several decades. A recent villain is the zebra mussel, which probably originated in the Caspian Sea and was likely first released into the Great Lakes from a ship's ballast water tank in the mid-1980s. Zebra mussels have now spread widely throughout the lakes and other waterways of eastern North America, where they have wreaked havoc with delicate ecological systems by ingesting large quantities of algae—a fundamental component of aquatic food webs. Zebra mussels also multiply rapidly, clogging water intake pipes and encrusting aquatic infrastructure and boats. The associated economic losses are enormous—they are expected to add up to a cumulative figure of at least $3.1 billion within the next few years.[5]

Terrestrial ecosystems are no less at risk. The damage wrought by the pesticide-resistant whitefly is a warning of the high stakes involved. The whitefly caused tens of billions of dollars of agricultural damage in California in the early 1990s before moving on to South America, where it has helped spread crop viruses that led to the abandonment of more than 1 million hectares of cropland. In the United States, the aggressive purple loosestrife plant has become a widely known symbol of the broader threat. It is thought to have first been accidentally introduced into North America in the late eighteenth century in wool imports and solid ship ballast, and then deliberately imported for ornamental and likely for medicinal purposes during the nineteenth century. Today, it has taken over more than 600,000 hectares of temperate and boreal wetland, crowding out native vegetation that is used by wildlife for food and shelter.[6]

The bioinvasion problem cries out for an international response. Among the steps that could be taken are inspections, limits on ballast water discharges, and the adoption of a precautionary approach that prohibits the knowing introduction of exotic species unless they have been shown to be benign. Some 23 different international treaties make at least some mention of exotic species, including the 1951 International Plant Protection Convention, the 1982 Law of the Sea, and the 1992 Convention on Biological Diversity. Although many of these agreements are quite weak, some of them include important commitments. The 1959 Antarctic Treaty, for one, banishes all exotics from the region unless they are specifically listed on an annex of exceptions or the bearer is granted an import permit. Besides legally binding treaties, a range of "soft law" instruments such as codes of conduct and action plans also address the bioinvasion threat.[7]

Tougher international agreements are needed to address this problem adequately, yet any accord stringent enough to alter today's rising tide of biotic mixing could run into conflicts with world trade rules. In what may be a foreshadowing of controversies to come, the Chinese government (which is not yet a member of the World Trade Organization, although it hopes to join soon) has complained that a ban imposed by the United States in late 1998 on the import of goods in untreated wooden packing crates amounts to an unfair trade barrier. The U.S. government imposed the ban after determining that Chinese packing crates were a primary culprit in the recent introduction of the voracious Asian long-horned beetle, an invasive insect that poses a major threat to the health of U.S. hardwood forests. The European Union recently placed similar restrictions on Chinese packaging, while China in turn limited the use of U.S. and Japanese crates made from coniferous trees after discovering wood-eating worms in some of them.[8]

TRADING IN WILDLIFE

Although habitat loss and the introduction of invasive species are the world's leading causes of diminishing biological diversity, for some species that are particularly prized on the international market—such as the tiger and the black rhinoceros—trade in the species itself is a major threat.[9]

The global trade in wildlife is a booming business. Each year, some 40,000 monkeys and other primates are shipped across international borders, along with some 2–5 million live birds, 3 million live farmed turtles, 2–3 million other live reptiles, 10–15 million raw reptile skins, 500–600 million ornamental fish, 1,000–2,000 raw tons of corals, 7–8 million cacti, and 9–10 million orchids. China, Europe, Japan, parts of Southeast Asia, and the United States are major consumers of wildlife and associated products for use as pets, in zoos, as clothing and ornamentation, and in medicine and horticulture. The wildlife trade is valued at some $10–20 billion annually, at least a quarter of which is thought to be illegal.[10]

Governments took an important step toward controlling the wildlife trade with the 1973 Convention on International Trade in Endangered Species of Wild Fauna and Flora (CITES), which 146 countries are now party to. This convention bans trade (with a few narrow exceptions) in more than 800 species in danger of extinction, such as the giant panda, Asian and African elephants, rhinos, sea turtles, and many species of monkeys, birds of prey, parrots, lizards, crocodiles, orchids, and cacti. Through a requirement for export permits, it also restricts trade in some 29,000 other species that are at risk of becoming threatened. This category includes hummingbirds, birds of paradise, black and hard corals, and birdwing butterflies. CITES is generally credited with substantially reducing trade in many threatened species, including gorillas, chimpanzees, cheetahs, leopards, and crocodiles.[11]

In one notable though controversial case, the members of CITES agreed to ban trade in ivory in 1990 in the face of rapidly declining elephant populations. Poaching declined dramatically after the ban, and many elephant populations have begun to recover. But several southern African states with good track records in elephant protection have long objected to the ban. These countries argue that limited and regulated ivory trade can in fact create an incentive to protect rather than decimate elephants, as it enables elephants to "pay their own way" in countries where elephants and impoverished peoples are often on a collision course.[12]

In response to pressure from these countries, CITES

members agreed in 1997 to allow for a limited one-time trade in stockpiled ivory between Japan and Botswana, Namibia, and Zimbabwe. This sale took place in the spring of 1999. It was intended as an experiment that, if successful, might pave the way for a broader resumption of controlled ivory commerce. But critics worry that allowing any trade at all will open a Pandora's box, fueling resumed poaching. Preliminary indications are that poaching is in fact on the rise. Elephant poaching in Kenya reportedly has increased fivefold since the ban was lifted. Zimbabwe has also experienced an upsurge in poaching, with at least 84 elephants slaughtered in 1999.[13]

Despite some notable CITES successes, unchecked trade in many threatened species continues apace. Nearly half of the world's turtle species face possible extinction, due in large part to the growing demand for turtles for both food and medicinal ingredients. In China, where turtles are thought to confer wisdom, health, or longevity, certain species now fetch up to $1,000 apiece. Many of the turtles sold in China actually originated in Viet Nam, Bangladesh, and Indonesia, as well as in the United States.[14]

The United States has become a major center of reptile trafficking in recent years—both legal and illegal. More than 2.5 million live reptiles were brought into the country in 1995, and in 1996 some 9.5 million reptiles were exported or re-exported, mainly to Europe and East Asia, according to estimates by TRAFFIC, a nonprofit wildlife trade monitoring group. Species such as the Komodo dragon lizard of Indonesia, the plowshare tortoise of northeast Madagascar, and the tuatara (a small lizard-like reptile from New Zealand) reportedly sell for as much as $30,000 each on the U.S. black market.[15]

On the other side of the world, Yemen is a major importer of African rhinoceros horn, which is sought after for use in the handles of traditional daggers known as *jam-*

biyas. TRAFFIC estimates that at least 75 kilograms of rhino horn were smuggled into Yemen each year from 1994 to 1996, and that horns from more than 22,000 rhinos may have been imported into the country since 1970. Fewer than 10,000 rhinos now remain in the wild in Africa, down from 70,000 in 1970.[16]

Wildlife trade is becoming a globally integrated industry, with air travel making it possible for popular pet species to be bred far from their native ranges, and then flown to pet stores on the other side of the world. When these exotics escape to the outside world, as they inevitably do, they can cause considerable damage to local ecological systems. Aquarium fish are one of the biggest culprits, but reptiles are also often complicit in biological pollution. A turtle known as the red-eared slider accounts for more than 80 percent of all reptile exports from the United States. Conservationists worry that turtles introduced for both pets and food will outcompete native turtle species in many parts of the world, particularly in East and Southeast Asia.[17]

MICROBES ACROSS BORDERS

In the first centuries of the Roman empire, growing commerce between Mediterranean civilizations and Asia precipitated the "great plague" of A.D. 165. Believed to have been smallpox, this epidemic claimed the lives of a quarter of the population of the Roman empire. In the fourteenth century, bubonic plague swept through Europe—the "Black Death." This epidemic, to which a third of Europe's population succumbed, was introduced into China as the Mongol empire expanded across central Asia, and from there spread by caravan routes to the Crimea and the Mediterranean.[18]

As the twenty-first century begins, the process of globalization is dramatically accelerating the pace at which microbes travel the globe. As the late AIDS researcher

Jonathan Mann of Harvard University explained, "The world has rapidly become much more vulnerable to the eruption and...to the widespread and even global spread of both new and old infectious diseases. This new and heightened vulnerability is not mysterious. The dramatic increase in worldwide movement of people, goods, and ideas is the driving force behind the globalization of disease." Only by looking out for the health of people everywhere is it now possible to promote healthy societies anywhere.[19]

The rapid growth in international air travel is a particularly potent force for global disease dissemination, as air travel makes it possible for people to reach the other side of the world in far less time than the incubation period for many ailments. At the same time, adventure tourism and other pursuits are drawing people to ever more remote locations, increasing the chance that microbes will be introduced to vulnerable populations.[20]

Environmental degradation is another powerful contributor to many of today's most pressing global health threats. The World Health Organization (WHO) estimates that nearly a quarter of the global burden of disease and injury is related to environmental disruption and decline. For certain diseases, the environmental contribution is far greater. Some 90 percent of diarrheal diseases such as cholera, which kill 3 million people a year altogether, result from contaminated water. And 90 percent of the 1.5–2.7 million deaths caused by malaria annually are linked with underlying environmental disruptions such as the colonization of rainforests and the construction of large open-water irrigation schemes, both of which increase human exposure to disease-carrying mosquitoes. A recent analysis by Cornell University ecologist David Pimentel and his colleagues reached an even starker conclusion—that some 40 percent of all deaths worldwide are attributable to environmental decline.[21]

When globalization and environmental decline join

forces, the health implications can be staggering. The power of this combination is demonstrated by the tragic history of the AIDS pandemic. As of 1999, the HIV virus had infected 50 million people worldwide, killing more than 16 million of them. In particularly hard hit countries in Africa, as much as a quarter of the population harbors the virus.[22]

The epidemic initially came to light at roughly the same time in the early 1980s in Africa, the Caribbean, and North America. The question of where the virus had originated was politically charged, with WHO skirting the issue for many years by maintaining that the virus had emerged simultaneously on at least three continents. "Few scientists accepted that position, recognizing it for what it was—a political compromise," notes author Laurie Garrett in her book *The Coming Plague.* "If humanity hoped to prevent its next great plague, it was vital to understand the origins of this one." In the last few years, scientists have made important strides toward getting to the bottom of this controversial question.[23]

It is now widely believed that HIV was originally harbored in chimpanzees inhabiting the West African rainforest, crossing over into human populations as early as the 1940s. Although exactly how this occurred will never be known, scientists speculate that it resulted from hunters cutting themselves while harvesting their kill, or perhaps through the direct consumption of raw meat. The epidemic thus may have had its origins in intermingling between humans and chimpanzees as a result of human incursion into previously remote forests. According to a theory put forth by Jaap Goudsmit of the University of Amsterdam, the decline in chimpanzee populations resulting from the human invasion might have created a biological imperative for the simian immunodeficiency viruses (SIV) to seek out new hosts—humans.[24]

Scientists believe that saving Africa's imperiled chimpanzees may be crucial for discovering a way to stave off the deadly HIV infection in humans, as the chimpanzees are immune from HIV's most lethal effects. But Africa's primates are under siege, with many on the verge of extinction. One major threat is the thriving "bushmeat" trade. As logging roads penetrate remote forests, loggers and hunters snare chimpanzees, gorillas, monkeys, bush pigs, snakes, and other prey. They either eat the meat themselves or transport it to West African cities, where bushmeat is considered a delicacy. "These chimps are information we need," warns Dr. Beatrice Hahn of the University of Alabama, who led a team that recently confirmed the link between AIDS and chimps. "Killing them for the pot is like burning a library full of books you haven't read yet."[25]

Another major outstanding question related to the origins of the AIDS epidemic is how HIV, once it was transferred from chimps to humans, made the leap from being an isolated condition confined to Africa's remote hinterlands to its current status as a global pandemic. Although many links in this chain are unknown, a range of phenomena are thought to have contributed, including warfare near the region from which the virus is thought to have first emerged; the paving of the TransAfrica highway, which provided an easy route for carrying HIV across the continent; population growth and urbanization; and, ultimately, burgeoning international travel and migration.[26]

As the movement of people into remote parts of West Africa's forests continues to pick up speed thanks to logging and hunting, scientists warn that other dangerous viruses may make the jump from primates to people. An even broader issue is at stake as well. "AIDS is trying to teach us a lesson," noted Jonathan Mann. "The lesson is that a health threat in any part of the world can rapidly become a health threat to many or all."[27]

Numerous other urgent global health challenges loom. Over the past two decades, more than 30 infectious diseases have been identified in humans for the first time, including AIDS, Ebola, Hantavirus, and hepatitis C and E. In a recent case that aroused widespread concern in the United States, health experts confirmed in October 1999 that at least five people in New York City and surrounding areas died from a new strain of the African West Nile virus, a rare mosquito-borne disease never before seen in the western hemisphere. They attribute the emergence of the disease to the steady rise in international trade and travel, concluding that the disease was transmitted either by a smuggled exotic bird or by an infected human who carried it into the country from abroad.[28]

Environmental disruption is also a potent contributor to today's microbial migrations. According to WHO, "environmental changes have contributed in one way or another to the appearance of most if not all" of the newly emerging diseases. Land use changes such as deforestation or the conversion of grasslands to agriculture that alter long-established equilibria between microbes and their hosts are sometimes to blame. In other cases, changes in human behavior are the culprit, such as careless disposal of food and beverage containers or car tires, which can create new breeding sites for disease-carrying organisms such as mosquitoes. Movements of pathogens themselves or the organisms that carry them are also sometimes the cause.[29]

An added problem is the reemergence of microbes thought to have been vanquished in some parts of the world. Cholera's reappearance in Latin America is a case in point. Until 1991, there had been no epidemic outbreaks of this deadly disease in this region for nearly a century. But the disease erupted with a vengeance in Peru that year, ultimately infecting some 322,000 people and killing at least 2,900 of them. The outbreak was catastrophic for the country's economy, causing importers to ban Peruvian fish and fruit from their markets, and tourists to avoid the country. All told, the economic costs to Peru's economy added up to $770 million—almost one fifth of normal export earnings. The outbreak quickly spread beyond Peru, contaminating the water supply of every country on the continent but Paraguay and Uruguay before it gradually wound down two years later. Across the Americas, the disease infected more than a million people and killed 11,000 during the first half of the 1990s.[30]

Scientists are trying to understand why cholera is now reemerging with such force. A number of factors seem to be at work. One theory is that the cholera bacteria was discharged from the ballast water of ships arriving in Peruvian ports from South Asia. Poor sanitation also undoubtedly played a major role, as cholera is often spread by contact with food or water that has been contaminated by human waste containing the bacteria. Another theory is that El Niño may have contributed to the outbreak by causing warmer ocean temperatures that encourage large blooms of plankton that can harbor the organism.[31]

If El Niño was in fact a key piece of the puzzle, then the cholera epidemic of the early 1990s was likely just a harbinger. Scientists project that climate change will lead to a surge in infectious diseases, both by increasing the range of disease-carrying organisms and by inducing a growing numbers of extreme weather events such as floods and hurricanes, which tend to leave epidemics in their wake. "There are strong indications that a disturbing change in disease patterns has begun and that global warming is contributing to them," notes Paul Epstein, Associate Director of Harvard Medical School's Center for Health and the Global Environment.[32]

Already, dengue fever and malaria both appear to be expanding their reach northward into cooler climates— locally contracted cases of malaria have been reported in recent years in Florida, Georgia, Texas, Virginia, New York, New Jersey, Michigan, and even Ontario. The record number of extreme weather events experienced in 1998 exacted a heavy toll on human health. Epstein reports that heavy flooding in East Africa led to large increases in the incidence of malaria, Rift Valley fever, and cholera; that delayed monsoons in Southeast Asia contributed to the region's wildfires, causing widespread respiratory ailments; and that Central American countries slammed by Hurricane Mitch experienced an increase in cholera, dengue fever, and malaria.[33]

Although the global interdependence of human and ecological health is creating frightening vulnerabilities, it is also generating an imperative for countries of the North and South to work together to confront shared perils.

Faced with raging transcontinental epidemics of cholera and plague in the mid-nineteenth century, European governments convened 12 International Sanitary Conferences between 1851 and World War I that forged a series of international health agreements covering issues such as quarantines, trade restrictions, and procedures for disease notification and inspection. In 1946, these and later efforts culminated in the creation of the World Health Organization, which has had a number of important successes in its first half-century, perhaps most notably the eradication of smallpox in 1977.[34]

This system provides a firm foundation on which to build the new biological controls needed to protect people and ecosystems from the introduction of disruptive exotic species and diseases. Although economic globalization dominated headlines at the close of the twentieth century, ecological integration may pose even greater challenges for international cooperation in the decades ahead.

CHAPTER 4

GLOBAL GROCERS

On New Year's Day 1994, a ragtag group of rebels calling themselves the Zapatista National Liberation Army took control of large areas of the impoverished Mexican state of Chiapas. Their armed rebellion was to protest a pattern of economic development that was enriching a few large landowners engaged in coffee production and ranching while denying the state's impoverished majority the land reform once promised by the country's constitution. It was no coincidence that the insurrection occurred on the same day that the North American Free Trade Agreement (NAFTA) entered into force. Among its many other effects, NAFTA was projected to put hundreds of thousands of Mexican peasant farmers out of business by undercutting them with cheaper, subsidized corn from the United States.[1]

The Chiapas uprising was indicative of broader insecurities playing themselves out around the world as a result of sweeping transformations under way in the world's agricultural markets. The last several decades have seen agriculture rapidly transformed in response to both technological change

and economic restructuring. As it becomes a globally integrated enterprise, farmers from poor countries find themselves in direct competition with the mechanized agribusiness
of the U.S. Corn Belt. In response to these pressures, traditional small farmers on every continent are rapidly being sup-
planted by large farms linked to the global marketplace.

Agriculture is integrally linked with the basic human
right to food security. It is also an important economic activity in most of the world—in low-income countries, agriculture accounts for an average of 30 percent of economic
output and over 60 percent of employment. But agriculture
is far more than an economic sector, and food is not just a
product like televisions and tires. Agriculture is also "a pillar of rural life, of the environment, of conserving old ways,"
as a Japanese trade negotiator recently put it. For this reason, today's agricultural upheavals have far-reaching implications for the welfare of billions of people as well as the
health of the natural world.[2]

THE AGRICULTURAL TRADE BALANCE

The value of world agricultural trade has soared in recent
decades, nearly doubling between 1972 and 1998 alone—
from $224 billion to $438 billion. (See Figure 4–1.) Agriculture accounts for 11 percent of the value of all world
exports. For some continents, this share is even higher—25
percent of Latin America's exports are agricultural, as are 18
percent of Africa's. Trade in basic food grains such as wheat,
rice, and corn dominates international agricultural exports
in volume terms, although nonessentials such as flowers,
coffee, and sugar dominate in value terms.[3]

Nearly 240 million tons of grain were exported in
1998—some 13 percent of total world production. But global aggregates mask great variations in export and import
dependence among countries and regions. Australia, for

FIGURE 4–1

World Exports of Agricultural Products, 1961–98

instance, exports nearly 63 percent of its grain production, and Japan imports 75 percent of its consumption. The U.N. Food and Agriculture Organization (FAO) has identified more than 80 "low-income food deficit countries," which it defines as poor countries that are net importers of food for at least three years in a row. More than half of these countries are in Africa. Collectively, they are home to the majority of the world's chronically undernourished people.[4]

Trade in food is intertwined with the course of human history. As far back as Roman times, grain imports from northern Africa helped sustain the empire. Europe became a net importer of grain during the Industrial Revolution in the nineteenth century, and the United States emerged as its primary breadbasket. By the middle of the twentieth century, North America was exporting 23 million net tons of grain per year, while Western Europe was importing 22 million tons. Asia was also beginning to rely on the grain trade, importing some 6 million net tons. Over the course of the next few decades, Australia and New Zealand emerged as important exporters, while imports surged in Asia, Eastern Europe and the former Soviet Union, Africa, and Latin America.[5]

During the 1980s, Europe's international grain position was turned on its head as its population stabilized and agricultural production surged in response to new technologies and generous governmental subsidies. For decades a major grain importer, by the early 1990s Western Europe was a net exporter of more than 20 million tons per year. (See Table 4–1.)[6]

The vast majority of internationally traded corn and soybeans is destined for the huge livestock feedlots of the industrial world, as well as for the smaller but growing livestock industries in developing countries. Direct trade in meat is also on the rise. Total meat exports—including beef, chicken, and pork—grew more than sixfold between 1961 and 1998, increasing from 3.5 million to 22 million tons. As with grain, industrial countries dominate the international meat trade: the United States, Canada, Australia, and the countries of the European Union between them account for 70 percent of the total. But a few developing countries are also significant exporters, including Brazil, China, and India.[7]

The developing world is a net importer of basic foodstuffs such as grain and meat, but it is a major exporter of many cash crops, such as bananas, coffee, cotton, soybeans, sugarcane, and tobacco. As of 1998, developing countries accounted for 95 percent of the exports of palm oil, 90 percent of cocoa, 88 percent of coffee, and 85 percent of bananas. Recent decades have seen particularly rapid growth in socalled nontraditional exports—principally flowers, fruits, and vegetables. These crops tend to command far higher prices than traditional agricultural exports, which have been in decline in recent decades. Exports of nontraditional crops

TABLE 4–1

The Changing Pattern of World Grain Trade, 1950–98[1]

Region	1950	1960	1970	1980	1990	1998
			(million tons)			
North America	+ 23	+ 39	+ 56	+125	+101	+ 86
Western Europe	– 22	– 25	– 28	– 7	+ 27	+ 19
Eastern Europe and Former Soviet Union	0	+ 3	0	– 45	– 29	+ 3
Latin America	+ 1	0	+ 6	+ 8	– 1	– 5
Africa	0	– 1	– 4	– 16	– 27	– 38
Asia and Middle East	– 6	– 17	– 33	– 63	– 71	– 81
Australia and New Zealand	+ 3	+ 8	+ 12	+ 12	+ 15	+ 21

[1]Plus sign indicates net exports; minus sign, net imports. Imports and exports do not balance out due to differences in export and import data and lags in shipment times.
SOURCE: Based on Lester R. Brown, *Who Will Feed China?* (New York: W.W. Norton & Company, 1995), derived and updated from U.S. Department of Agriculture, Economic Research Service, *Production, Supply, and Distribution*, electronic database, Washington, DC, November 1994 and December 1999.

from Central America increased in value by an average of 17 percent annually between 1985 and 1992, and exports of these crops from South America (excluding Brazil) increased at 48 percent a year over this period. Chile has pursued this export path with particular abandon. Chilean fruit exports—including table grapes, apples, pears, peaches, avocados, citrus fruits, berries, and melons—rose 16-fold in value between 1994 and 1997 alone. By 1997, they were bringing in $1.6 billion—10 percent of total export earnings.[8]

The average distance that food travels as it makes its way from farm to table has climbed steadily in recent decades as agriculture has become globally integrated. A study by the London-based National Food Alliance found that food consumed in the United Kingdom on average traveled more than 50 percent further over the last two decades. A small jar of strawberry yogurt eaten in Germany has components that travel more than 3,000 kilometers, according to a report by the Wuppertal Institute. Though the milk is available locally, the strawberries are grown in Poland, and the packaging materials come from Austria and Switzerland.[9]

In the new century, water scarcity will increasingly shape the pattern of the world grain trade. Water shortages have become a major constraint on agricultural productivity in many regions of the world in recent decades, as human numbers have climbed and as agriculture has become increasingly water-intensive owing to the widespread adoption of fertilizer-intensive, high-yielding varieties. Because water itself is difficult and expensive to transport long distances, countries facing water shortages generally import grain rather than water. "With each ton of grain representing about 1,000 tons of water, countries in effect balance their water books by purchasing grain from other countries rather than growing it themselves," explains Sandra Postel of the Global Water Policy Project.[10]

This grain-for-water strategy is workable so long as enough countries have surpluses available to export, and so long as the grain-importing countries have enough foreign exchange to pay the bill. But the large number of countries that are expected to become net grain importers over the next several decades owing to water scarcity, growing populations, and other factors may undermine this assumption. Postel projects that the number of people living in countries where water is sufficiently scarce to necessitate grain imports will climb from about 470 million today to more than 3 billion by 2025, most of whom will live in highly impoverished countries in Africa and South Asia. Growing world demand could cause food prices to spike, exacerbating social pressures in impoverished food-importing nations.[11]

Despite the difficulty of transporting water internationally, a number of proposals are afoot for ambitious efforts to ship large quantities of water across international borders. In one controversial recent case, the Canadian province of Ontario granted a permit in 1998 to a company called the Nova Group to transport some 600 million liters of Lake Superior water to Asia in bulk tankers. After a storm of protest, the provincial environment ministry revoked the permit. Continued pressure from private companies who want to enter the water export business has led the U.S.-Canadian International Joint Commission to convene public hearings exploring the implications of possible bulk water exports from the Great Lakes.[12]

Concern is rising among Canadian activists that provisions of NAFTA and of the World Trade Organization (WTO) will impede their efforts to restrict large-scale water export schemes. Already, the California-based Sun Belt Inc. has sued Canada under NAFTA, claiming it is entitled to $10.5 billion in compensation for a ban on water exports imposed by British Columbia several years ago. Sun Belt's Canadian partner was one of six companies granted export licenses during a drought in California in the 1980s.[13]

COUNTING ENVIRONMENTAL AND SOCIAL COSTS

Agricultural exports have helped fill foreign-exchange coffers in developing countries, but they have also imposed heavy social and environmental costs. As governments and international lending institutions promote cash crops at the expense of subsistence agriculture, women and the poor often lose out because of their relative lack of access to land, credit, and other resources. And social structures can be severely disrupted as farming communities are broken apart to service large and distant export markets.[14]

The clearing of land for export-oriented cash crops is a major cause of deforestation. Wildfires in Indonesia in recent years were sparked by fires deliberately set to clear land for palm oil and pulpwood plantations. Palm oil exports from Indonesia more than doubled between 1991 and 1997, climbing from 1.4 million to nearly 3 million tons. With the encouragement of the International Monetary Fund, the Indonesian government is planning to boost exports further in the years ahead as part of its strategy for climbing out of economic crisis. (See Chapter 8.) On the other side of the globe, in the Amazon basin, efforts to boost soybean exports have set in motion plans to construct an extensive network of canals, highways, and railroads in order to get the crop to market—principally in Asia and Europe. Environmentalists worry that these projects will fuel further deforestation of the region's unique and diverse ecosystems.[15]

Cash crops are often grown with heavy doses of pesticides, imperiling the health of both agricultural workers and food consumers. Nontraditional exports such as flowers and fruit are doused with particularly high doses of toxic pesticides, in part to meet importers' desires for "blemish-free" produce. The flower industry is the most lethal of all for workers, as flowers are not ingested and are thus not subject to food safety requirements. A study of nearly 9,000 workers at Colombian flower plantations found exposure to 127 different pesticides. Some 20 percent of the pesticides used on these plantations are either banned or unregistered in the United Kingdom or the United States. Two thirds of Colombian flower workers report suffering headaches, nausea, impaired vision, and other symptoms as a result of pesticide exposure, according to the Colombian Human Rights Committee in Washington, DC. Colombia has surpassed California as the principal supplier of roses, carnations, chrysanthemums, and

other flowers to the U.S. market, accounting for two thirds of
all fresh-cut flowers sold there today.[16]

Beyond its tragic toll on the health of farm workers, pesticide dependence also poses health risks for importing
countries and economic risks for exporters. Between 1984
and 1994, the U.S. Food and Drug Administration detained
more than 14,000 shipments of fruits and vegetables from
10 Latin American and Caribbean countries because of
excessive pesticide residues, according to a World Resources
Institute analysis. The associated economic losses to the
exporting countries added up to $95 million.[17]

The global trade in meat also causes environmental
destruction. In Central America, the lure of the export
market for beef spurred a massive clearing of the rainforest
for cattle ranching during the 1960s and 1970s, a period
when the region was exporting large amounts of beef
to North America to satisfy the U.S. appetite for hamburgers
and steak. In Botswana, heavy dependence on beef
exports has resulted in land degradation from overgrazing;
half of the country's beef production is exported, much of
it bound for the European market. And in Somalia, research
by ecologist Bruce Byers concluded that rapid growth
in exports of sheep, goats, and cattle over the last several
decades contributed to a tragic breakdown of the country's
traditional, ecologically balanced nomadic system of
livestock rearing. The result has been overgrazing, soil
erosion, and the degradation of rangelands, all of which
diminish the ability of the land to provide sustenance for
the Somali people.[18]

FISHERIES AT RISK

Just as small farmers around the world are threatened by the
integration of agricultural markets, many fishing communities are struggling under strains imposed in part by global-

ization. More than 200 million people around the world are economically dependent on fishing in one way or another. And nearly 1 billion people worldwide, most of them in Asia, rely on fish as their primary source of animal protein. But the world's fisheries are under siege as a result of habitat destruction, pollution, and overexploitation, with 11 of the world's 15 major fishing grounds and 70 percent of the primary fish species either fully or overexploited.[19]

The lure of international markets is a driving force behind this growing crisis. Fish exports have climbed nearly fivefold in value since 1970, reaching $52 billion in 1997. (See Figure 4–2.) By volume, nearly half the fish caught today are traded, up from only 32 percent in 1970. Industrial countries dominate global fish consumption, accounting for more than 80 percent of all imports by value. Developing countries, on the other hand, contribute nearly half of all exports. Their share of the total has risen steadily in recent

FIGURE 4–2

World Fish Exports, 1970–97

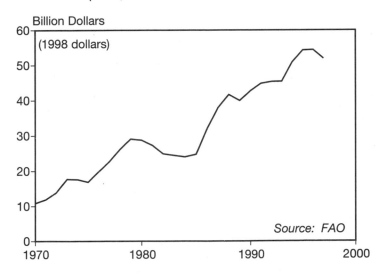

decades as fleets have turned south in search of fish in
response to the overfishing of northern waters. In 1970,
developing countries accounted for 37 percent of all fish
exports, measured by value; by 1997, their share had risen
to 49 percent. Chile, China, Indonesia, and Thailand are all
now among the world's top 10 fish exporters. (See Table
4–2.) Exports from these four countries alone quadrupled in
value between 1980 and 1997.[20]

The 1982 Law of the Sea treaty granted coastal states the
right to control resource development within 200-mile
Exclusive Economic Zones off their shores. Subsequent
access agreements have spelled out the terms by which foreign fleets could ply distant water. Ships from Europe, Japan,
Russia, South Korea, and Taiwan have entered into agreements with African countries to fish for tuna, hake, octopus,
squid, and shrimp, many of which are now fully or overexploited. The African countries involved have often granted

TABLE 4–2

Top 10 Fish Exporters and Importers, 1997

Country	Exports (billion dollars)	Share of World Total (percent)
Norway	3.4	7
China	2.9	6
United States	2.9	6
Denmark	2.6	5
Thailand	2.3	5
Canada	2.3	4
Chile	1.8	3
Taiwan	1.8	3
Indonesia	1.6	3
Spain	1.5	3
World	51.0	100

access to their fisheries for cut-rate prices, and the people
and fishing communities most affected by these agreements
have generally been left out of the negotiations and received
little of benefit. Similar agreements stipulate the terms of
access for fleets from the United States, Taiwan, Japan, and
South Korea to the rich tuna fisheries of Pacific island
nations such as the Solomon Islands and Kiribati.[21]

With many Third World fisheries now becoming depleted as well, overfishing for export markets means depriving
small-scale fishers of their catch. Demand from foreign markets also drives up the price of domestically available fish to
the point where they are beyond the means of local people.
In Senegal, for instance, species once commonly eaten
throughout the country are now either exported or available
only to the elite.[22]

Exports of farmed fish have proved particularly lucrative
for many countries over the last few years—but also partic-

TABLE 4–2 *(continued)*

Country	Imports	Share of World Total
	(billion dollars)	(percent)
Japan	16.0	28
United States	8.1	14
Spain	3.1	5
France	3.1	5
Italy	2.6	5
Germany	2.4	4
United Kingdom	2.1	4
Hong Kong	2.1	4
Denmark	1.5	3
China	1.2	2
World	56.0	100

SOURCE: U.N. Food and Agriculture Organization, *Fishery Statistics Yearbook*,
vol. 85 (Rome: 1999).

ularly costly ecologically and socially. One in five fish produced today comes from a farm. Although aquaculture has
the potential to alleviate pressure on natural fish stocks, the
industry also has a number of liabilities of its own. It is often
land- and water-intensive. And many of the higher-value
farmed species, such as shrimp and salmon, are themselves
carnivores, which means that large numbers of lower-value
fish are sacrificed to feed them. For each kilogram of farmed
salmon and shrimp, 5 kilograms of wild oceanic fish are harvested and ground into high-protein pellets.[23]

Shrimp aquaculture has grown particularly fast in many
developing countries over the last few decades. In Thailand,
shrimp and prawn production surged from 61,000 tons in
1970 to a peak of 389,000 tons in 1995, with exports
accounting for 60 percent of the 1995 total. Nearly 70
percent of the country's shrimp harvest in 1995 was farmed
rather than caught from the sea. Thai shrimp production has
since declined as a result of the Asian financial crisis, though
it is expected to rebound as the economy recovers. Other
countries where shrimp aquaculture has taken off include
Bangladesh, Ecuador, India, Indonesia, Malaysia, and
Viet Nam. The revenues are substantial: in 1997, shrimp
exports accounted for nearly 16 percent of Ecuador's total
export earnings. But the repercussions are also grave,
including coastal pollution, the displacement of local people
from their land, and the clearing of large tracts of coastal
mangrove forests.[24]

Once regarded as wastelands, mangrove ecosystems are
now recognized as playing a critical role in protecting coast-
lines and serving as spawning grounds for oceanic fisheries,
thereby providing sustenance for local people. Yet they are
rapidly being felled to make way for shrimp farms. More
than a million hectares of mangrove forests have been lost to
fish farms over the last decade. In Thailand alone, some
253,000 hectares of the country's original 380,000 hectares

of mangrove forests have already been destroyed by shrimp farms, according to the country's National Economic and Social Development Board.[25]

FROM GREEN REVOLUTION TO "GENE" REVOLUTION?

Over the ages, farmers have relied upon diverse crop varieties as protection from pests, blights, and other forms of crop failure. Where traditional agriculture is still practiced, farmers often have extensive knowledge of the attributes of diverse varieties: the Ifugao people on the island of Luzon in the Philippines can name more than 200 varieties of sweet potato, for instance. Modern agriculture still depends on this rich storehouse of biological knowledge, as plant breeders and genetic engineers turn to traditional varieties for the genetic raw material needed to increase yields and produce seeds with attributes such as pest or disease resistance.[26]

The last century has seen a steady erosion of genetic diversity in agriculture as farmers have gradually replaced traditional varieties with more uniform crops. FAO estimates that 75 percent of crop genetic diversity has been lost over this century. This process accelerated in the 1960s with the widespread introduction of high-yielding Green Revolution varieties in many parts of the world. In the United States, more than 70 percent of all cornfields are now planted in just six varieties of corn. In India, farmers grew as many as 30,000 varieties of rice 50 years ago; today, three fourths of India's rice fields are planted with fewer than 10 varieties. And in Mexico, only 20 percent of the corn varieties that were cultivated in the 1930s can still be found today. The rapid pace at which plant genetic diversity is disappearing is leaving the world vulnerable to multibillion-dollar crop losses and reducing the storehouse from which future agricultural strains can be derived.[27]

Just as the Green Revolution transformed the practice of agriculture worldwide in the 1960s, the world may now be on the verge of a "Gene Revolution." Transgenic seeds (those that include genes transplanted from other species) have been in the research pipeline for decades, but it is only within the last few years that they have begun to be widely commercialized. As of 1999, some 40 million hectares of cropland worldwide had already been planted with transgenic varieties, more than triple the area they covered in 1997, and more than 20 times as much as in 1996. The United States dominates these statistics, accounting for 72 percent of the total acreage. Still, 11 nations besides the United States already have at least some land dedicated to transgenics. In much of the rest of the world, widespread public concern about the health and ecological impacts of eating and growing bioengineered crops has slowed their adoption. The area planted in transgenic seeds may level off over the next few years, as farmers try to gauge to what extent public concern and government regulation will cut into the global market for this food. (See Chapter 7.)[28]

Proponents of using genetic engineering in agriculture argue that it can be harnessed to wean farmers from their dependence on chemicals by producing plant varieties that are pest- and disease-resistant. They also envision the development of salt- and drought-resistant varieties that might permit production on marginal lands, as well as the creation of even higher-yielding varieties than those produced by the Green Revolution. Yet skeptics worry that the new herbicide-resistant varieties will entrench rather than reduce reliance on chemicals. They also worry about broader ecological disruption as bioengineered traits are accidentally but unavoidably passed on to neighboring plants through cross-pollination.[29]

The seed varieties of the Green Revolution were generated by public-sector research institutions and made freely available to farmers and researchers to be adapted to local circumstances and needs. The use of these seeds was thus consistent with the millennia-old agricultural practice of seed-saving, whereby farmers save and replant their seeds from year to year, gradually selecting the hardiest, best-adapted strains. Today's Gene Revolution, in contrast, is commercially driven and defined by patent rights. Biotechnology companies have successfully lobbied for increasingly strong—and increasingly global—patent protection, in some cases making it illegal for farmers to save and replant seeds. Prohibitions against replanting seeds and efforts to develop a technology that can prevent harvested seed from germinating have raised a storm of protest, particularly in the developing world, where more than 80 percent of all crops are grown from saved seeds.[30]

Recent years have seen biotechnology firms lay claim to a wide range of plant varieties though patenting. Governments have granted patents on transgenic soybeans, cotton, and rice varieties, as well as on traditional crops such as the Neem tree, which has long been cultivated for medicinal and agricultural uses in India, and on quinoa, a high-protein cereal eaten by millions of indigenous people in the Andes. The quest for patents has set off a wave of consolidation in the biotechnology and seed industry over the last few years, as industry giants such as Monsanto, DuPont, and Novartis have bought up other companies and their patents.[31]

Many people in the developing world view the patenting of indigenous knowledge for commercial gain as a form of theft—or "bio-piracy." They argue that if anyone deserves compensation for protecting and perfecting seeds, it is the farmers who have cultivated and selected them over thousands of years. These "farmers' rights" were implicitly affirmed in the 1992 U.N. Convention on Biological Diversity, as well as in earlier declarations negotiated under the auspices of FAO. But the 1993 Uruguay Round of trade talks that created the World Trade Organization paid little heed to these earlier agreements. Rather, heavy industry pressure led to requirements that countries pass legislation to bestow intellectual property protection on plant breeders and biotechnology corporations, while providing no such protection to farmers.[32]

THE WTO AND FOOD SECURITY

Because of agriculture's social and cultural importance, countries have historically been hesitant to bring the sector within the bounds of world trade rules. But the Uruguay Round of trade negotiations subjected the sector to trade disciplines for the first time.[33]

The impetus for the WTO's foray into agriculture dates back to the 1980s, when the European Union and the United States both began to pour billions of dollars into agricultural export subsidies in a bid to outcompete each other in overseas markets. Besides draining national treasuries, these export subsidies and related domestic agricultural policies spurred intensive agricultural techniques, which caused overproduction and associated environmental stress. "Driven by these incentives, farmers adopt chemical-intensive monocultures that lead to more soil erosion, chemical runoff, loss of biological diversity, and conversion of once-natural ecosystems to cropland than would otherwise have taken place," argues economist Robert Repetto.[34]

The subsidy-induced agricultural surpluses also had adverse social repercussions, particularly in the developing world. Excess supplies of basic commodities such as corn and wheat caused world prices to stagnate. The depressed prices harmed other agricultural exporters who could not afford to compete in the agricultural subsidy "arms race" between Europe and the United States. Argentina's export earnings, for instance, fell by 40 percent between 1980 and 1987 due to falling world prices for cereals and oilseeds. Farmers serving local rather than export markets also suffered, as cheap imported grain flooded their markets.[35]

By taking aim at agricultural export subsidies, the Uruguay Round's agricultural agreement could have been a net plus for both the environment and poor farmers in the developing world. But in the end, power politics prevailed. The European Union and the United States agreed to only minor subsidy reductions that they were able to largely avoid through the choice of the baseline year and the use of other escape hatches. The agreement allowed industrial countries to continue to support their farmers by converting export subsidies into direct income payments. But it required developing countries to phase down the agricultural import restrictions that are their primary tool for protecting domestic farmers from being forced out of business by subsidized imported grain. The results threaten to devastate millions of poor farmers. "For us, the price we get for yellow corn is a matter of life and death. It shapes our lives, our health and our future," explains Rosa Laranjo, a farmer from the island of Mindanao in the heart of the Philippines' "corn basket."[36]

Given these high stakes, it is hardly surprising that agriculture was a contentious issue at the Seattle WTO meeting. The Marrakesh accord of 1994 that created the WTO called for new negotiations on agriculture at the end of 1999. The United States and other agricultural exporters, including many from the developing world, are pushing hard for more access to overseas markets. But some countries that protect their farmers from international competition, including Japan and South Korea, worry that cheap imported grain will put their own farmers out of business. Walden Bello of the Bangkok-based Focus on the Global South warns that further agricultural liberalization "will drive the [Asian] region's small farmers over the edge."[37]

CULTIVATING GREENER GARDENS

The current direction of global agribusiness is ecologically unsustainable and socially disruptive. But farmers around the world are experimenting with a range of alternatives that could be scaled up to form the basis for a more sustainable future.

One way to counter the adverse effects of agricultural globalization is to support local agriculture. Many people are doing just that, as evidenced by a renaissance of urban gardening as well as the growing popularity of farmers' markets and other forms of community-based agriculture. Nonetheless, there remains a role for international agricultural commerce—if it can be radically overhauled.[38]

Particularly promising are alternative forms of agricultural trade that generate revenue for impoverished countries and communities while at the same time promoting environmental sustainability and social equity. This approach was pioneered in the 1970s by the "fair trade" movement, which promotes trade in goods that conform to social criteria, including adequate working conditions and a price for producers that compensates for stagnant commodity prices, and which ensures that profits are not lost to middlemen. The Netherlands' Max Havelaar Foundation was an early pioneer, launching a brand of "fair trade" coffee from Mexico in 1988. Imports of this coffee into the Netherlands rose steeply, climbing from just 253 kilos in 1988 to more than 3,000 by the late 1990s. The coffee is now produced by 200 cooperatives in 18 different countries in Africa and Latin America. It is sold in 90 percent of Dutch supermarkets and is widely available in several other European countries, including Belgium, Denmark, France, Germany, and Switzerland. A similar coffee, TransFair, can be found in eight nations, including Austria, Canada, Germany, Italy, Japan, and the United States.[39]

These fair trade coffee initiatives include general requirements aimed at encouraging environmentally sound cultivation. But a range of other initiatives now stipulate far more
specific environmental commitments. Organically certified
coffee is in growing demand worldwide, and several Central
and Latin American countries already have significant
amounts of land dedicated to producing it, including El Sal-
vador, Guatemala, Mexico, and Peru. Coffee certified as
"shade-grown" has also recently soared in popularity in
northern markets, driven by growing awareness that traditional "shade-grown" coffee plantations—an important
source of habitat for threatened bird populations—are rapidly being replaced in many areas by sun-grown, intensive
coffee cultivation. Several U.S.-based environmental organizations, including the American Birding Association, Conservation International, and the National Audubon Society, are
now promoting signature brands of shade-grown coffees.[40]

International trade in environmentally friendlier commodities has moved far beyond coffee. Spurred by growing
consumer demand for food that is both healthy and environmentally sound, organic agriculture has become a
growth industry in many parts of the world. The overall
global market for organic food is now worth an estimated
$11 billion annually, and is predicted to increase to $100 billion over the next 10 years, with most of the growth in the
United States, Europe, and Japan. Organic sales in North
America more than quadrupled between 1990 and 1998,
when they reached some $4.7 billion. U.S. organic farmers
have found a thriving market for their wares in Japan, where
an estimated 3–5 million consumers regularly purchase
organic produce.[41]

Many developing countries have moved to tap into the
international organic market, beckoned by consumers willing to pay "green premiums" that can run as high as 50–200
percent above regular prices. Mexico has been particularly

quick to leap at this opportunity. It now has some 10,000 organic farms on 15,000 hectares of land, most of them run by small farmers. In addition to their mainstay, organic coffee, Mexican farmers cultivate a range of other organic products as well, including apples, avocados, coconuts, cardamom, honey, and potatoes.[42]

Argentina has also rapidly moved into organic production—since 1992, the area devoted to organic farming has increased nearly 50-fold, reaching almost 231,000 hectares in 1997 (although organic production still only accounts for less than 1 percent of the country's total agricultural output). Sales of organic items such as fresh fruits, beef, milk, cheese, chicken, and olive oil rose from $1.5 million in the early 1990s to $20 million by mid-decade, and were expected to surpass $100 million by 2000. Some 74 percent of the organic production is exported, with nearly 83 percent of it going to Europe, 17 percent to the United States, and less than 1 percent to Japan.[43]

Certification programs are proliferating to help spur a transition to environmental and social sustainability in global agricultural markets. The New York–based Rainforest Alliance runs a Better Banana Project that gives certified bananas an ECO-OK label. To earn the mark, producer plantations must agree, among other things, not to clear any virgin forests and to monitor rivers and wells for pesticide residues. More than a quarter of all banana production in Costa Rica is now from certified lands, as is 41 percent of Panama's. The program works with both small and large growers. In a major victory, Chiquita Brands has now certified all company-owned farms in Costa Rica and is working to cover the rest—including farms in Panama, Colombia, Guatemala, and Honduras—by early 2000. The program has been criticized in some quarters, however, for paying insufficient attention to social concerns. A competing Fair Trade Banana was launched in Europe in 1996 by the Max Havelaar Foundation. Despite the controversy, the Rainforest Alliance is also setting itself up to certify producers of other commodities, including coffee, cacao, and oranges.[44]

Similar efforts are afoot to transform the world's seafood markets. In 1996, the World Wide Fund for Nature teamed up with one of the world's largest seafood product manufacturers, Anglo-Dutch Unilever, to create a Marine Stewardship Council (MSC)—modeled on the Forest Stewardship Council—to devise criteria for sustainable fish harvesting. The MSC is now an independent organization whose members include fishers' organizations, fish processors and buyers, food retailers, environmental groups, governments, and business leaders. The MSC symbol is expected to make its debut on packaged fish in early 2000. Unilever, which buys 25 percent of the world's white fish every year, has pledged not to buy any fish products after 2005 that are not certified as sustainably harvested.[45]

One looming worry is that these certification programs could be challenged as trade barriers at the World Trade Organization, because they distinguish between products based on how they were produced. WTO rules generally frown on such distinctions. (See Chapter 7.) To promote a transformation of world agriculture, governments will need to not only protect certifications from WTO challenges, but also actively support them through government procurement programs and other initiatives.

A range of other reforms are also needed if world trade is to support rather than undermine sustainable agriculture, including redirecting remaining agricultural subsidies in support of small-scale, low-input producers. Farmers' groups from around the world were out in force at the WTO meeting in Seattle, with many of them pushing for policies to promote environmental sustainability and social cohesion in the world's agricultural markets.[46]

The failure to find common ground on agriculture in

Seattle contributed to the overall breakdown of the talks.
Governments are scheduled to revisit agricultural issues in
WTO negotiations over the next few years. The acrimony in
Seattle suggests that agriculture will be a divisive issue on
the world stage for some time to come.[47]

CHAPTER 5

THE EXPORT OF HAZARD

On the night of December 2, 1984, a storage tank at a pesticide plant in Bhopal, India, owned in part by the U.S.-based Union Carbide corporation burst open, sending a cloud of poisonous methyl isocyanate gas toward the Jayaprakash Nagar shantytown that bordered the plant, and from there on to the rest of the city. "Slowly, the people of Bhopal in India's Hindi-speaking heartland began to awaken to horror and death," writes former *New York Times* correspondent Sanjoy Hazarika. "The city began to cough, to choke and heave, as tens of thousands woke to a suffocating, acrid white-yellow mist....Then the panic began as people saw husbands, wives, parents and children struck down—gasping for breath, clutching at burning, hurting eyes and chests, frothing at the mouth...and then choking on their own vomit and blood." The accident would claim more than 6,000 lives within a week and over 16,000 to date, going down in history as one of the world's worst environmental disasters.[1]

Due to a globalized world economy, developing countries

are trying to cope with thousands of hazardous industrial chemicals they did not invent, and that they have little capacity to regulate adequately. Although the chemicals have a range of important economic uses, Bhopal shows the Faustian bargain they often represent.

The only silver lining to the tragedy was the international spotlight it placed on chemical hazards and on the multinational companies sometimes implicated in generating them. Sparked by the horror of Bhopal and other industrial catastrophes, the world community has made some progress over the last few decades in crafting international rules to govern the commerce in hazardous wastes, products, and industries. But gaping holes remain in the global safety net.

TOXIC TRADE

A few years after Bhopal, world attention was once again focused on international toxic threats when a string of high-profile attempts to export hazardous waste received widespread publicity. Waste disposal costs were soaring at that time in many industrial countries in response to tighter regulations as well as shortages of landfill capacity, prompting several efforts to ship waste to poor developing countries desperate for cash.

In one notorious incident, the U.S. city of Philadelphia decided to solve a problem it was having by loading toxic ash from its municipal incinerators onto a ship called the *Khian Sea*, which set sail in August 1986 searching for someone to take the waste. But the strategy backfired. The ship initially toured the Caribbean Sea for a year and a half trying to find a country willing to accept its load. It finally dumped some of the wastes on a Haitian beach, provoking an uproar. It pulled up anchor once again, and after touring five continents and changing its name three times, the ship finally discharged the rest of its load in an undisclosed location in late 1988, according to its owners. Greenpeace, which has played a leading role in monitoring and exposing the waste trade, suspects that the ash was eventually dumped in the Indian Ocean in November of 1988.[2]

That same year, the small Nigerian fishing village of Koko found itself in the international spotlight when 8,000 drums of highly toxic waste, including methyl melamine, dimethyl formaldehyde, ethylacetate formaldehyde, and about 150 tons of polychlorinated biphenyls (PCBs) in the backyard of villager Sunday Nana began leaking. Visitors to the site described drums "popping from the sun" and smoking, while acid fumes reportedly engulfed the village. But the villagers were ignorant of the dangers. "The odor comes to my compound. It is everywhere," Nana told the *African Concord*, a weekly Nigerian newspaper. "But, to be sincere, it has not worried my health. I even walk in some places with bare feet. My children do the same." An Italian waste disposal firm was eventually held responsible, and the waste was returned to Italy. But the damage had already been done. Many of the Nigerian workers who helped remove the waste were hospitalized with severe chemical burns, nausea, vomiting of blood, and partial paralysis, and one person fell into a coma. Two years later, Nana himself passed away, although the Nigerian government claimed that he succumbed to a respiratory failure unrelated to the dumping.[3]

These shocking incidents spurred the international community to action. Many developing countries unilaterally banned waste imports at around that time: 33 countries had done so by 1988. And the U.N. Environment Programme (UNEP) accelerated negotiations toward an international agreement to regulate the waste trade. In 1989, the Basel Convention on hazardous waste export was finalized, requiring exporters to notify the recipient nation of a shipment and to receive approval for it before proceeding. But many observers found this little cause for celebration, contending that the accord legitimized a trade that should have been banned outright. In its early years, the treaty appeared to do little to stem the waste trade tide. Greenpeace estimated that more than 2.6 million tons of hazardous wastes were shipped from industrial countries to the South or the East between 1989 and 1994. Over roughly this same period, at least 299 dumpings were documented in Eastern Europe, 239 in Asia, 148 in Latin America, and 30 in Africa.[4]

As pressure mounted to strengthen the accord, the number of countries unilaterally deciding to ban waste imports climbed steadily—more than 100 had done so by 1994. Finally, the Basel Convention was itself strengthened in March of that year to ban all waste exports from industrial to developing countries—a victory for the South and a decision that Greenpeace hailed as "The Pride of the Basel Convention." But the ban will only have legal force when the 1994 amendment has been ratified by 62 countries. So far, only 17 countries have taken this step, although most are already respecting its terms. Still, a few key governments continue to object to some provisions of the ban, including Australia, Canada, New Zealand, and the United States.[5]

Despite the progress made in recent years in controlling the hazardous wastes trade, the problem remains far from solved. UNEP estimates that some 440 million tons of hazardous wastes are generated worldwide every year, about 10 percent of which is shipped across international borders. Illegal trade is believed to be flourishing, although it is impossible to quantify, given that most such trade never comes to light. "We presume that organized crime is behind it....We fear we don't hear too much about it because it is much like the illegal arms or drug trade," says Pierre Portas of the Basel Convention secretariat. Nonetheless, officials have intercepted illegal shipments originating in the United States that were bound for Ecuador, Guinea, Haiti, Malaysia,

Mexico, Panama, and Sri Lanka.[6]

In one horrifying case, in December 1998 nearly 3,000 tons of mercury-contaminated concrete waste produced in Taiwan by the Formosa Plastics Corporation were dumped in plastic shipping bags with no warning labels in a field outside the Cambodian port city of Sihanoukville. Unsuspecting local people initially scavenged among the materials, believing the bags could be used for floor mats or tarpaulins, and that the crushed concrete might become fertilizer. A worker who had unloaded the material from the ship died after suffering symptoms consistent with mercury poisoning, as did a villager who slept on one of the shipping bags. As local people became aware that the material was toxic, riots erupted among citizens angry at the officials believed to be responsible, and more than 10,000 residents fled the city in fear.[7]

Ironically, the growing strength of the environmental movement in Taiwan was partly to blame for the situation. In a replay of the situation in the United States and other industrial countries in the 1980s, growing environmental awareness in many rapidly industrializing Asian countries in the 1990s has made it difficult to dispose of waste domestically, creating a strong incentive for companies to look overseas. After the uproar in Cambodia, Formosa Plastics eventually agreed to remove the waste. It had initially planned to ship it to the United States for disposal, but the U.S. Environmental Protection Agency (EPA) decided to reconsider its initial approval after activists charged that plans to dispose of the waste in a low-income, predominantly Latino community in southern California amounted to environmental racism. Community concerns also thwarted subsequent efforts to dump the waste in the states of Idaho, Nevada, and Texas. As of November 1999, the mercury-tainted waste was still impounded at Taiwan's Kaohsiung Port, after having also been rejected by France.[8]

CIRCLES OF POISON

Ever since the publication of Rachel Carson's *Silent Spring* in 1963, concern over the effects of toxic chemicals on the health of both people and wildlife has been a driving concern of the environmental movement. But only in recent years have we begun to understand the ease with which hazardous chemicals move across international borders, catapulting the issue from the national to the global plane.

One of the most vivid demonstrations of the global reach of today's chemical world comes from the Arctic. Research over the last few decades has revealed that some persistent organic chemicals travel thousands of miles from their source before reaching this remote part of the world. These chemicals "evaporate from soils as far away as the tropics, ride the winds north, then condense out in the cold air of the Arctic as toxic snow or rain," explains Fred Pearce of the *New Scientist*. Certain harmful chemicals are particularly likely to follow this route, including PCBs, hexachlorocyclohexane, toxaphene, and chlordane. Scientists believe that these chemicals can circle the globe at a rapid rate, traveling as far as from India to the Arctic in as little as five days.[9]

The long-range transport of hazardous substances leads to the ironic result that people and wildlife in some of the world's most remote places are being exposed to some of the highest levels of chemical contamination anywhere on Earth. PCBs as well as a range of harmful pesticides have built up in the Arctic food chain, reaching ever higher concentrations further up in the chain—for example, from fish to seals to polar bears to whales and ultimately to people.[10]

Researchers began to understand this phenomenon in the mid-1980s, when scientists were looking for a control sample of breast milk from Inuit women in the Canadian Arctic. They had assumed that this milk would be completely pure, and thus useful to compare with the breast milk of women living in the midst of the industrial heartland. Instead, the researchers were stunned to measure PCB levels in the milk of Inuit women from Broughton Island in northeastern Canada that were the highest ever found in any human population except those who had been exposed to industrial accidents. Subsequent research has revealed similar contamination in many other parts of the vast Arctic.[11]

Besides wind and water currents, international commerce is another potent mechanism through which hazardous chemicals move about the world. The pesticides trade is a case in point. Over the last 50 years, pesticide use has surged more than 50-fold, increasing from 50 million kilograms a year in 1945 to some 2.7 billion kilograms a year today. And today's pesticides are more than 10 times as toxic as those used in the 1950s. Exports of pesticides have surged nearly ninefold since 1961, reaching $11.4 billion in 1998. (See Figure 5–1.) Since the 1970s, growing awareness of the dangers of pesticides has led over 90 countries to ban the domestic use of various compounds, including the "dirty dozen"—particularly harmful pesticides such as DDT, endrin, and chlordane. Yet in a phenomenon that has come to be known as the "circle of poison," banned pesticides that are exported to other countries sometimes return to their country of origin on imported food.[12]

Developing countries are particularly vulnerable to health and environmental damage from pesticides, as many of them lack the regulatory mechanisms needed to evaluate risks thoroughly or to ensure that chemicals are used according to instructions. Protective gear for pesticides is often not worn, as it is not suitable for tropical climates. And warning labels on imported pesticides are often non-existent, vague, or written in languages that farm workers cannot read. The World Health Organization (WHO) estimates that these deficiencies cause some 25 million agricultural workers in the developing world to suffer at least one inci-

FIGURE 5–1

World Pesticide Exports, 1961–98

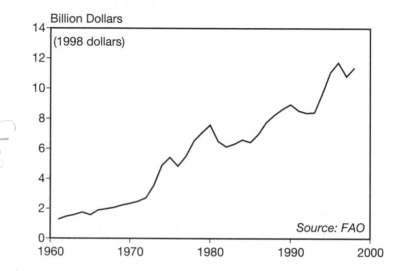

dent of pesticide poisoning per year, resulting in as many as 20,000 deaths annually.[13]

In a number of documented cases over the years, pesticides exported from industrial countries have been implicated in health disasters in the developing world. Use of the extremely toxic pesticide DBCP has been linked with sterility in more than 26,000 farm workers in Costa Rica, the Philippines, and 10 other countries. Costa Rica, for one, imported 5 million kilograms of DBCP from the United States between 1966 and 1973—more than 2 kilograms per citizen—for use on banana plantations either owned by Del Monte, Dole, or Chiquita or producing bananas for them.[14]

In 1994, a group of farm workers from these 12 countries filed a class action suit in U.S. courts seeking damages from the U.S. companies that produced or used DBCP after it was already known to have caused sterility among U.S. farm workers. (The chemical was banned for use in the United

States in 1979.) Five of six defendant companies have settled the case, agreeing to pay $52 million in damages, while admitting no liability. Dole Fresh Fruit Co. is the sole remaining defendant. Barry Levy, an adjunct professor with Tufts University's Department of Family Medicine and Community Health, warns that the DBCP debacle "may be just the 'tip of the iceberg' of a series of such catastrophes" around the world.[15]

Despite widespread attention to the "circle of poison" phenomenon, the export of pesticides not approved for domestic use is rising rather than falling—at least from the United States. The Foundation for Advancements in Science and Education (FASE), a Los Angeles–based public interest communications and research group, recently conducted an in-depth review of U.S. pesticide exports. They found that more than 312 million kilograms of pesticides were exported from U.S. ports in 1996—a 40-percent increase since 1992. A substantial share of these were shipped without identification of specific chemical names in the public shipping documents, making it difficult to quantify what proportion were either banned or restricted for domestic use. Nonetheless, FASE conservatively estimates that at least 10 million kilograms of such pesticides were exported from the country in 1995 and 1996. Perhaps even more alarming, 13 million kilograms of pesticides that WHO classifies as "extremely hazardous" to agricultural workers were exported in 1996, a more than fivefold increase over 1992 levels.[16]

U.S. exports of pesticides that are thought to disrupt the endocrine system, which regulates the secretion of hormones into the bloodstream, are also on the rise. Suspected endocrine disrupters exported from the United States in significant quantities in recent years include alachlor, chlordane, heptochlor, and metribuzin. Nearly 33 million kilograms of pesticides in this category were exported from U.S. ports in 1996—an average rate of some 100 tons per day. This marked a 28-percent increase over the average daily exports of such chemicals over the previous four years. Most of the shipments were bound for Argentina, Belgium, Brazil, India, Japan, and the Philippines.[17]

Many countries now find themselves saddled with growing stocks of obsolete and unused pesticides. The U.N. Food and Agriculture Organization (FAO) estimates that several hundred thousand tons of obsolete pesticides are piling up worldwide, with more than 100,000 tons in the developing world. A sizable share of the chemicals were originally donated under foreign aid programs. They include highly toxic and persistent chemicals such as aldrin, DDT, dieldrin, lindant, and parathion. The pesticides can no longer be used, either because they have deteriorated while in storage, because they have by now been banned, or because the country no longer needs them.[18]

In many cases, the chemicals are being stored in hazardous conditions. For instance, drums are often out in the open, where exposure to sunlight and rain can cause them to leak and corrode. In some areas, they are being stored near markets, where they are easily accessible to children. They are also contaminating soils, groundwater, irrigation, and drinking water. FAO estimates that it would cost $80–100 million in Africa alone to dispose of the accumulated stocks adequately.[19]

In addition to pesticides, numerous other dangerous products are disseminated through trade—sometimes because of their hazardous properties. In Canada, for instance, when domestic sales of asbestos declined due to public health concerns, the industry collaborated with the government to promote sales abroad. Canada is now the world's leading exporter and second-largest producer of white asbestos. Ninety-six percent of the asbestos Canada produces is now exported—the majority of it to the Third World. Worried that a growing number of bans on asbestos production in Europe will devastate its asbestos industry, Canada has tried to protect its market by turning to the World Trade Organization (WTO), where it has lodged a complaint against France, arguing that France's 1996 ban on the production of white asbestos breaks international trade rules because it was imposed with insufficient scientific evidence of adverse health effects. The European Union as a whole has now followed France's lead in banning white asbestos, which may ignite an even larger WTO showdown.[20]

In an effort to crack down on the export of hazardous materials, more than 50 countries gathered in Rotterdam in September 1998 to finalize an international treaty that puts in place a system of prior informed consent for trade in 22 pesticides and 5 industrial chemicals when these substances are banned or restricted in the exporting country. This accord builds on an earlier FAO nonbinding code of conduct on the distribution and use of pesticides. Negotiations now under way on a convention on persistent organic pollutants are aimed at banning altogether 12 particularly hazardous chemicals.[21]

POLLUTION HAVENS?

Not only is trade in hazardous products thriving, but recent decades have also seen hazardous industries themselves become widely dispersed around the planet. In many cases, these industries are becoming concentrated in the developing world, where safety practices and environmental enforcement and monitoring are often rudimentary at best.

The asbestos industry is a case in point. Asbestos production has plunged in most industrial countries over the last 25 years as evidence has accumulated that breathing asbestos fibers causes lung cancer. But production and use continue to climb in many countries, including Brazil, China, India, Indonesia, Poland, South Africa, South Korea, and Thailand. In Brazil, domestic asbestos consumption is increasing 7 percent annually. The country also exports some 70,000 tons of asbestos per year, principally to Angola, Argentina, India, Mexico, Nigeria, Thailand, and Uruguay. The surge in asbestos use in the developing world is expected to cause anywhere from 30,000 to several million deaths over the next 30 years.[22]

Although they sound innocuous enough, many "recycling" operations in developing countries also pose grave environmental dangers. For example, millions of used car batteries are sent from the United States every year to smelters in Brazil, China, India, Japan, Mexico, and South Africa, among other countries, to be melted down for lead recovery. But the smelting process exposes workers to dangerous lead levels, causing classic symptoms of lead poisoning, including headaches, dizziness, stomach cramps, and kidney pains. Excessive exposure to lead can cause more serious long-term health problems, including kidney damage, reproductive problems, and brain impairment in children. Scrap recovery businesses based on imported materials often cause similar contamination problems.[23]

Over the last few decades, the developing world has become home to a growing share of the hazard-laden petrochemical industry. In 1980, 11 percent of all chemicals were produced in developing countries; by 1996, this figure had grown to 18 percent. Much of this expansion involves joint ventures with multinational firms. For example, the chemical industry's share of total U.S. foreign direct investment (FDI) in manufacturing in developing countries increased from 18 to 34 percent between 1990 and 1998. Approximately 41 percent of U.S. FDI in the Philippines in 1998 was in chemicals, as was 22 percent of such investment in Colombia.[24]

High-tech industries such as computers and electronics have also gone global in recent years. And despite their early reputation as relatively clean, these high-tech industries often exact heavy environmental costs. Semiconductor manufacturing, in particular, is a toxic-laden business. The manufacturing process employs hundreds of chemicals, including arsenic, benzene, and chromium, all of which are known carcinogens. California's Silicon Valley is a testament to the industry's dangers: it is home to 29 sites listed on EPA's Superfund list of the country's most contaminated toxic dumps, giving it the dubious distinction of hosting the largest concentration of such sites in the country.[25]

As high-tech industry spreads around the world, it is bringing its environmental liabilities along with it. The industry has grown particularly rapidly in Southeast Asia. In the Philippines, for example, exports of electronics equipment surged from just over $1 billion in 1985 to above $10 billion in 1996—more than half of the country's total export earnings. Semiconductors accounted for nearly 80 percent of this sum. A 1996 review of 22 computer-related companies based in industrial countries by the Silicon Valley Toxics Coalition of San Jose, California, found that more than half of the manufacturing and assembly operations—processes intensive in their use of acids, solvents, and toxic gases—are now located in developing countries.[26]

A debate has raged over the years about the extent to which industries might be fleeing tightening environmental regulations in industrial countries by seeking out "pollution havens" in the developing world. Studies suggest that industries are generally drawn to overseas locations by a range of factors, including the cost and quality of labor, the availability of natural resources, and the access to large markets. In most cases, environmental control costs alone are not high enough to be a determining factor in location decisions. But even if companies move to the developing world for other reasons, they may well take advantage of lax environmental laws and enforcement once there.[27]

In a few instances, moreover, relaxed environmental
enforcement does appear to have been a motivating factor in
companies' location decisions. The controversy over the
1993 North American Free Trade Agreement (NAFTA) put
the spotlight on one notoriously polluted region where this
seems to have been the case for some firms—the border
between northern Mexico and the United States. That area is
the site of some 3,200 mostly foreign-owned manufacturing
plants known as *maquiladoras*. In the city of Mexicali, near
the California border, more than a quarter of the factory
operators surveyed in the late 1980s said that Mexico's lax
environmental enforcement influenced their decision to
locate there.[28]

These and other companies helped make the area an
environmental disaster zone. A survey conducted by the
U.S. National Toxics Campaign in the early 1990s found
toxic discharges at three quarters of the *maquiladoras* sam-
pled. Chemicals known to cause cancer, birth defects, and
brain damage were being emptied into open ditches that ran
through the shantytowns around the factories. High rates of
severe birth defects and other health problems have been
detected along the border. Particularly horrifying have been
elevated rates of spina bifida, a spinal-nerve defect, and
anencephaly, a fatal condition in which babies are born with
incomplete or missing brains, in the heavily polluted area
that straddles Brownsville in Texas and Matamoros in Mexico. Despite the environmental side agreement that accompanied NAFTA, conditions have improved little and may even
have deteriorated in the years since, as more U.S. companies
have flocked to the region.[29]

The *maquiladoras* region is but one of some 850 export
or special processing zones worldwide that collectively
employ some 27 million workers. These zones normally permit goods to be imported duty-free, on the condition that
they then be used to produce exported products. A range of

other inducements encourage companies to locate production in these zones, including tax holidays and free land or reduced rent. There is considerable evidence that one lure is often a casual attitude toward substandard labor practices, such as dangerous working conditions and restrictions on the right to organize.[30]

Although no comprehensive data on the question have been gathered, environmental abuses are undoubtedly also common. In the coastal Cavite province near Manila, for instance, local fishers accuse Taiwanese- and Korean-owned factories located in the special economic zones adjacent to Manila Bay of dumping pollutants that are responsible for killing thousands of fish. And the Chinese National Environmental Protection Agency has accused firms from Taiwan and South Korea of setting up shop in China in order to flee tougher environmental regulations at home.[31]

A CLEANER PATH

A few decades ago, developing countries often argued that pollution was the price of progress. But the last several years have brought an environmental awakening to most corners of the globe. Environmental laws and enforcement are gradually being strengthened in response.

Rather than setting themselves up as pollution havens, many developing countries are recognizing that they have an opportunity to learn from the mistakes of the industrial world, and leapfrog directly to the technologies of tomorrow. Such products and processes will be far cleaner and more efficient in their use of energy and raw materials than the equipment typically in use today—and thus far healthier for the people that use or live among them.

International investment can help expedite this transition, as many companies bring advanced technology with them when they undertake new investments abroad. The lure of selling into "greener" international markets can also have a salutary impact on environmental performance in the developing world. Shi Yonghai, a Senior Researcher at the Chinese Academy of International Trade and Economic Cooperation, maintains that "if Chinese traders don't pay attention to environmental protection and ecology when producing or purchasing goods for export, it will be impossible for China's export sector to grow, or even to maintain its current levels."[32]

Limited evidence also suggests that the recent move of many governments to privatize state-owned factories by selling them to domestic or foreign private investors sometimes promotes cleaner industrial processes. One reason is that privatization eliminates the conflict of interest that arises when the government is both producer and regulator. In addition, the pressure to turn a profit introduces an incentive to adopt manufacturing techniques that reduce energy and materials use and thus diminish pollution.[33]

While international markets have often spread environmental horrors, they can also be harnessed on behalf of the transition to cleaner technologies that use resources efficiently and produce little if any hazardous waste. But stronger environmental rules of the road will be needed if the globalization process is to support this shift.

CHAPTER 6

SHARING THE AIR

In May 1985, a team of British scientists stunned the world with an article in *Nature* magazine that reported a remarkable 40-percent loss of stratospheric ozone over Antarctica between September and October 1984. Despite extensive research on the subject, no such precipitous decline had been predicted by the atmospheric models the scientists relied on. Indeed, the ozone losses were so unexpected that the investigators at first suspected instrument error and delayed the release of the data. But subsequent satellite readings confirmed the presence of this massive ozone "hole"— which covered an area the size of the continental United States. The findings revealed that during the Antarctic spring, ozone levels were becoming low enough to present serious risk of cancer, cataracts, and other health problems in New Zealand and other southern countries.[1]

When the "ozone hole" revelations hit the headlines, international negotiations aimed at limiting the use of chlorofluorocarbons (CFCs), the chemicals suspected of thinning Earth's protective ozone layer, were well under way—but badly stalemated. Scientists had warned for years that without international action, depletion of the ozone layer would increase the intensity of ultraviolet (UV) radiation and cause millions of additional skin cancer cases, sharply diminish agricultural yields, and kill aquatic organisms. But industry leaders had persuaded governments that the cost of replacing the chemicals was too high.[2]

As the negotiations approached a crucial stage, news of the massive hole in the ozone layer—accompanied by dramatic, computer-generated color images—provided clear evidence that ozone depletion was a more unpredictable and dangerous phenomenon than most scientists thought. This turned ozone depletion from a scientific abstraction to a tangible threat, profoundly altering the atmosphere of the talks. The accumulating scientific evidence of ozone depletion also spurred industries that produce and use the CFCs that cause ozone depletion to accelerate their research into practical, affordable alternatives. Some even realized that those who moved fastest to the new chemicals might actually gain market share as the older chemicals were phased out by international agreement.[3]

Suddenly the plodding negotiations turned into an avalanche of key decisions. Just over two years after the discovery of the ozone hole, on September 16, 1987, negotiators meeting in Montreal finalized a landmark in international environmental diplomacy: the Montreal Protocol on Substances That Deplete the Ozone Layer. This treaty mandated far-reaching restrictions in the use of CFCs as well as halons, another group of ozone-damaging chemicals.[4]

Ozone depletion is a quintessentially global problem: CFCs released mainly in northern industrial countries are destroying a protective layer of the atmosphere nearly everywhere—and doing so most dramatically in the remotest and supposedly unpolluted "upper" and "lower" corners of the world. But ozone depletion is global for another reason: the technologies that cause it are a twentieth-century invention that spread rapidly around the world as a result of the acceleration of global trade and investment that marked the final decades of the century. The response to ozone depletion has also been global, with diplomats around the world—advised by scientists, and lobbied by businesses and environmental organizations from dozens of countries—breaking new ground in international law and diplomacy in order to turn the problem around.

The leading role of international scientists, the constructive efforts of multinational companies, the concerted pressure of environmental groups from a range of countries—and the dramatic results—all suggest that the Montreal Protocol was a high point for environmental globalization. But the biggest test is likely to come with another atmospheric problem, and an even more global issue: climate change.

For 12 years, government leaders from around the world have struggled to forge an effective international agreement to slow the emission of carbon dioxide (CO_2) and other greenhouse gases that are steadily building in the atmosphere, thanks in large measure to the fossil fuels that powered the twentieth century. Although a first effort at a climate convention was ratified in the early 1990s, and the subsequent Kyoto Protocol received preliminary approval in 1997, the world is still far from an agreement that has the far-reaching effects of the Montreal Protocol—and even further away from a real solution to the problem. Resistance to change by leading industries and political bickering among key governments are preventing the kind of commitment that is needed to solve this most global of problems.[5]

Even as globalization has helped fuel the unprecedented buildup of greenhouse gases in recent decades, so must a new approach to globalization be realized if that growth is to be slowed. In their efforts to overcome this impasse in the next few years, the world's governments will reveal much about whether the positive potential of globalization can overcome its negative effects in the early decades of the twenty-first century.

MEDDLING WITH THE ATMOSPHERE

The successful conclusion of the ozone treaty negotiations in Montreal was widely hailed at the time as a historic event. The protocol was the most ambitious attempt ever to combat environmental degradation on an international scale. Governments from poor countries as well as rich, from the East as well as the West, were involved in the talks. The protocol they agreed on would have extensive effects on the multibillion-dollar global industry that produced the offending chemicals, as well as on the numerous businesses that manufactured products dependent on them, such as the rapidly growing computer chip industry. Billions of consumers also faced changes in products they had grown accustomed to, such as foam coffee cups and car air conditioners. The accord was signed on the spot by 24 nations and the European Community, and has since been ratified by more than 170 countries.[6]

In the years since the Montreal meeting, the accord has been strengthened several times to require deeper emissions cuts and coverage of more chemicals. It has succeeded in setting in motion myriad responses by national governments, international organizations, scientists, private enterprises, and individual consumers—with decisive results. January 1, 1996, was an important milestone, as the protocol required CFC production for domestic use in industrial countries to be phased out altogether by then. By 1997, global production of the most significant ozone-depleting substance—CFCs—was down 87 percent from its 1987 level. (See Figure 6–1.)[7]

FIGURE 6–1

World Production of Chlorofluorocarbons, 1950–97

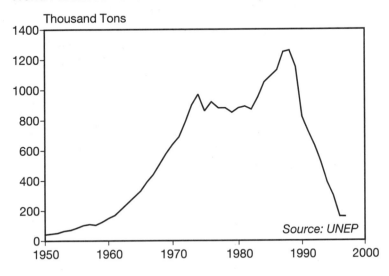

Despite the encouraging decline in CFC production, the world is currently suffering through the period in which the ozone layer will be most severely damaged. This is due to the long time lag between when CFCs and other ozone-depleting compounds are released and when they reach the stratosphere. And once there, CFCs can persist for centuries. The largest "ozone holes" on record have developed above the Antarctic over the last few years. Ozone losses over mid to high latitudes in both the northern and southern hemispheres have also increased rapidly, leading to higher levels of UV radiation over populated and agriculturally productive corners of Earth, such as Canada, Chile, and Russia.[8]

The increased levels of UV radiation reaching Earth are thought to be having the expected range of adverse effects on human and ecological health, including impaired immune systems, elevated skin cancer rates, and disruption of aquatic ecosystems. Current estimates suggest that if all countries comply with the Montreal Protocol, the ozone shield will gradually begin to heal within the next few years, but a full recovery to pre-1980 levels is not expected until about 2050.[9]

The climate change treaty, in contrast, is not yet strong enough to put a meaningful dent in the buildup of greenhouse gases in the atmosphere. Global emissions of carbon dioxide, the most important of these, have increased nearly fourfold since 1950, and CO_2 concentrations in the atmosphere are more than 30 percent above preindustrial levels, reaching their highest level in 160,000 years. If the world continues on its current fossil-fuel-intensive course, scientists estimate that CO_2 concentrations will double by the year 2100, increasing the average temperature at Earth's surface by 1.0–3.5 degrees Celsius (3–8 degrees Fahrenheit).[10]

Even assuming that the commitments made in December 1997 at Kyoto are fully implemented, they are only projected to slow the buildup of CO_2 concentrations modestly. Under the Kyoto accord, industrial countries agreed to collectively reduce their emissions of greenhouse gases to 6–8 percent below 1990 levels between 2008 and 2012. Because emissions from developing countries are not yet limited by the accord, projections suggest that global CO_2 levels could reach as much as 30 percent above 1990 levels over the next 15 years, even with the commitments currently stipulated in the accord. Yet scientists estimate that emission cuts on the order of 60–80 percent below current levels will likely be required to eventually stabilize CO_2 concentrations in the atmosphere.[11]

Most climate scientists believe that global warming has begun. The Intergovernmental Panel on Climate Change (IPCC), a group of 2,500 scientists from around the world that advises the climate negotiators—and modeled on the international scientific process used to guide the ozone negotiations—reported in 1995 that "a discernible human influence on global climate" could already be detected. Record-breaking temperatures over the last several years have supported this conclusion: 14 of the warmest years since recordkeeping began have occurred since 1980. The temperature increase in 1998 was particularly pronounced, making that year the warmest on record. (See Figure 6–2.)[12]

Although it is difficult to predict the precise effects of climate change, the international scientific community has warned that they can be expected to be serious. In its 1995 assessment, the IPCC warned that a doubling of CO_2 concentrations would cause enough warming to raise sea levels by between 15 and 95 centimeters over the next century. The resulting flooding of coastal areas would turn millions of people into environmental refugees in low-lying areas of the world, and cause several island nations to disappear altogether. The IPCC also predicted that doubled CO_2 concentrations would cause a dramatic increase in extreme weather

FIGURE 6–2

Global Average Temperature at Earth's Surface, 1866–1998

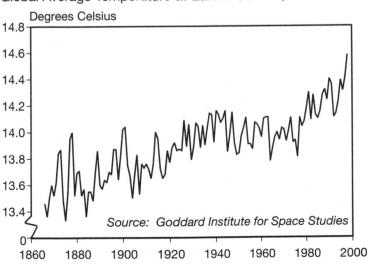

Source: Goddard Institute for Space Studies

events such as storms and hurricanes, disrupt ecosystems worldwide, and precipitate a surge in the transmission of infectious diseases such as malaria. Climate change is expected to exacerbate water scarcity in arid regions such as the Middle East, and to diminish agricultural productivity in many of the world's poorest countries. One recent study found that climate change–induced drought could increase the share of Africa's population at risk of hunger by as much as 18 percent by the 2050s.[13]

SHOWDOWN IN KYOTO

The world community took a tentative first step toward confronting climate change at the June 1992 Earth Summit, when the U.N. Framework Convention on Climate Change was finalized. The treaty's deliberately ambiguous language urges but does not require industrial nations to hold total emissions of greenhouse gases to 1990 levels or below by 2000. In addition, all signatories, including developing countries, are obligated to conduct emissions inventories, submit detailed reports of national actions taken to implement the convention, and work to take climate change into account throughout their social, economic, and environmental policies.[14]

Within a few years, it became clear that the Convention urgently needed strengthening, just as happened in the case of ozone depletion, when the original 1985 Vienna Convention was rapidly superseded by scientific, technological, and political developments. Few industrial countries were on track to meet the greenhouse gas stabilization target. Even if they had been, further efforts would be needed to achieve the treaty's broader aim, which is to stabilize atmospheric concentrations at a level that will prevent "dangerous... interference with the climate system."[15]

It was against this backdrop that negotiators arrived in Kyoto in December 1997. Most industrial countries appeared ready to accept legally binding reduction targets. But it was not yet clear how large they would be, and how they would be distributed among nations. Another controversial issue cast a long shadow over Kyoto: whether and how developing countries should participate in the agreement. Two years earlier, discussions of developing-country commitments had been explicitly excluded from the negotiating mandate for Kyoto. It was generally agreed that industrial countries should be the first to take on reduction commitments, as per capita carbon emissions levels in these countries are on average six times higher than in the developing world. But the U.S. Senate passed a unanimous resolution prior to Kyoto saying it would not ratify any agreement that did not contain binding targets for developing countries. The U.S. government thus arrived in Kyoto determined to secure commitments from the developing world.[16]

It looked as though the stage had been set for stalemate. But on the final day of the conference, negotiators struck a last-minute deal mandating the 6–8 percent reduction from 1990 levels by 2008–12. On the surface, this seemed to be a significant step forward. Disaster in Kyoto appeared to have been averted, and the protocol was widely hailed as historic. But it quickly become clear that celebration was premature. The accord papered over serious differences among countries, and left critical details still to be resolved.[17]

Although the headlines out of Kyoto focused on the 6–8 percent reduction goal, the devil lay in the details. In particular, a crucial Annex A detailed how the collective emissions goal was to be shared among industrial nations. For most countries, the goal hovered around 8 percent. But there were notable exceptions, such as an 8-percent increase in emissions granted to Australia. Further complicating the situation was the variation in emissions trends since the 1990 base year. Owing to economic collapse, Russia's 1995 emissions were 29 percent below 1990 levels, and Ukraine's 1997 emissions were down by 49 percent. And in 1997 the European Union's emissions were 4 percent below while U.S. emissions were 11 percent above 1990 levels.[18]

These discrepancies in national targets took on particular significance because the protocol created a path-breaking yet controversial emissions trading scheme that allows countries and companies to purchase emissions credits from one another. Countries such as Russia and Ukraine, which under the protocol can now increase their emissions substantially (see Table 6–1), as they are already well below the 1990 level, were by the stroke of the pen granted carbon emissions credits potentially worth billions of dollars annually. Critics pointed out that this system was creating what they dubbed "hot air"—marketable emissions rights that were the result of reductions already achieved.[19]

The "hot air" option allows major emitters such as the United States to meet most of their reduction commitment by purchasing credits from abroad for reductions that have already taken place. Emissions trading is based on sound economic theory, providing an incentive for reducing emissions where it can be done most cost-effectively. But trading of "hot air" would undermine the legitimacy of this system, making it more an arena for political horse-trading than a market mechanism. An added problem is the lack of institutions at the international level to conduct the monitoring and verification needed to make a complex emissions trading system work.[20]

Besides the uncertainties over emissions trading, another shadow hovering over Kyoto was the failure to reach agreement on the question of how to involve developing countries in the treaty. This led the U.S. government to announce that it would not submit the accord for Senate ratification until it could convince key developing countries to agree to "meaningful participation" in the protocol. As of late 1999, only 18

TABLE 6–1

Greenhouse Gas Commitments and Emissions Trading
Potential for Selected Countries Under the Kyoto Protocol[1]

Country	Emissions, 1997[2]	Emissions Commitment for 2008–12[3]		Trading Potential as of 1997[4]
	(million tons of carbon equivalent)	(percent change from base year)		(million tons of carbon equivalent)
Industrial Countries				
Australia	121	122	+ 8	+ 1
Canada	186	154	− 6	− 33
Eur. Union[5]	945	904	− 8	− 42
Japan	349	311	− 6	− 38
Norway	15	14	+ 1	− 1
United States	1807	1516	− 7	−291
Countries in Transition				
Bulgaria	23	34	− 8	+ 11
Czech Rep.	43	48	− 8	+ 5
Estonia	6	10	− 8	+ 4
Latvia	4	9	− 8	+ 5
Poland	116	145	− 6	+ 28
Romania	45	66	− 8	+ 22
Russian Fed.	586	828	0	+243
Ukraine	127	250	0	+123

[1]Includes emissions of the six gases designated in the Kyoto Protocol: CO_2, CH_4, N_2O, HFCs, PFCs, and SF_6. Excludes emissions removal from land use change and forestry. [2]Data for Spain and Russian Federation are for 1995; data for Australia, Belgium, Greece, and Romania are for 1996. [3]Emissions commitments are here expressed in relation to the base year of 1990; under the protocol, however, some Parties actually use earlier base years: Bulgaria (1988), Poland (1988), and Romania (1989). [4]Plus sign indicates emissions available for potential export, minus sign indicates potential emissions imports. [5]European Union (EU) data exclude Italy, Luxembourg, and Portugal. The emissions commitments for the 15 EU countries were initially listed as –8 for each, but subsequently revised under the "bubbling" provision of Article 4, which groups these countries as one and allows them to redistribute their individual emissions commitments in ways that preserve the collective total. SOURCE: See endnote 18.

countries had agreed to be bound by the protocol, far fewer than the 55 required to give it legal force.[21]

ATMOSPHERIC POLITICS

During the run-up to Kyoto in 1997, television viewers throughout the United States were subjected to a steady barrage of advertisements paid for by an organization with an innocuous name—the Global Climate Information Project. But protecting Earth's climate is low on the priority list of this group, whose members include industries and unions that feel far more threatened by action to combat climate change than by climate change itself. Among the members of the coalition are the American Petroleum Institute, the American Plastics Council, the National Mining Association, and the AFL-CIO, a federation of labor unions representing some 13 million workers.[22]

In an ominous tone, the ads warned that the pending Kyoto accord would inflict enormous economic pain on the U.S. economy. They then professed to clue viewers in on a shocking fact—that U.S. economic competitors in China and other developing countries would not be included in the Kyoto targets. (Not mentioned was the fact that the average American is responsible for nearly eight times as much carbon per capita as the average Chinese.) The punch line was designed to stick: "This UN Treaty Isn't Global and It Won't Work."[23]

At the same time that the advertisements were running in the United States, Lee R. Raymond, Chairman of Exxon, gave a speech in China on behalf of the American Petroleum Institute in which he questioned whether climate change was a real problem and urged developing countries to increase rather than limit their fossil fuel consumption. Raymond also issued a veiled threat: "Competition among countries eager to develop petroleum reserves is at an all-time high." He then advised developing countries to offer tax concessions as well as "rational environmental standards" in order to attract foreign investment.[24]

These two arguments provide an unusually potent prescription for political impasse—on the one hand, convince the U.S. Congress to block the Kyoto Protocol unless developing countries also adopt carbon targets, then convince these same countries that targets would hinder their development. The result: ensuring that ratification of any accord to emerge from Kyoto would be tied up for years to come. The strategy was devious and the advertisements highly misleading. But the process underscored an important political reality: climate change is inextricably linked with broader insecurities about economic welfare in a global age. Until these anxieties are addressed head-on, there will be little hope of ensuring climate stability in the twenty-first century.

Concerns about international competitiveness are hardly new in environmental diplomacy. They were a prominent feature of early ozone diplomacy. But in the ozone case, worries about competitiveness were effectively turned on their head and used as an argument for cooperative international action rather than an excuse for intransigence.

In the mid-1980s, it appeared increasingly likely that the United States was going to move forward with stringent domestic restrictions on ozone-depleting substances. This pressure for action was driven by several factors. High on the list was growing public concern about the health impacts of ozone depletion—in particular, the projected growth in skin cancer rates. The companies that produced CFCs worried that they might find themselves with both tarnished public images and mounting legal liabilities. CFC producers and other affected industries were also concerned about an ongoing lawsuit in which the Natural Resources Defense Council was suing the Environmental Protection Agency (EPA) for not implementing provisions of the U.S. Clean Air Act that required the agency to take action on ozone depletion. U.S. businesses feared being held to a tougher standard than their international competitors. In order to avoid this outcome, the industry-backed Alliance for Responsible CFC Policy did an abrupt about-face in late 1986 and began to advocate international controls on CFCs.[25]

European industry was slower to come around, partly because of less public attention to the issue in Europe, which made companies there uninclined to view regulation as inevitable. Furthermore, some Europeans apparently believed that research into substitutes was more advanced in the United States, which they feared would put European firms in a vulnerable position in the battle for dominance of CFC-substitute markets. Eventually, however, the European calculus shifted. One impetus may have been the pending U.S. ozone legislation, which would have imposed trade restrictions on the imports of countries that had not undertaken comparable domestic action. The Europeans also apparently realized that it would be easier for them than for the United States to implement the production targets being proposed for the Montreal meeting, as European countries had not yet eliminated the use of CFCs in aerosols—a sector for which alternatives were both cheap and readily available. Armed with this knowledge, the Europeans changed course and eventually supported decisive action to reduce and eventually eliminate the use of CFCs.[26]

With climate change—which results from the actions of hundreds of different industries as well as billions of consumers—the industrial politics are more complex. Large, powerful industries such as coal, oil, chemicals, steel, and automobiles continue to sponsor the kind of misleading ads described earlier, while many other companies and industries are beginning to take the kind of constructive approach that ultimately drove the ozone negotiations to success. As with ozone depletion, these companies argue that the effort to replace fossil fuels with new energy technologies will create at least as many economic opportunities as it threatens. But so far, these voices of industrial reason are being drowned out by those who fear they will be losers in the race to reduce greenhouse gas emissions.[27]

In a transatlantic about-face from the earlier situation with ozone, some European industrialists are skittish about the European Union getting out in front of the United States on climate change, as they fear their competitiveness will be harmed if European companies are required to make investments in climate-friendly technologies while their U.S. competitors are not.[28]

Today's controversies about the participation of developing countries in the climate change accord also have antecedents in ozone history. When the Montreal Protocol was first negotiated, developing countries used only small quantities of CFCs. Yet their consumption was projected to grow rapidly in the years ahead as they strove to raise living standards by providing refrigerators, air conditioning, and other amenities. If these countries did not participate in the accord, growth in developing-world CFC consumption would likely soon swamp any reductions in industrial countries. China and India were of particular concern. Though neither was at the time a significant CFC consumer, together they accounted for nearly 40 percent of the world's population— and both had plans to increase dramatically the production of consumer goods that could contain CFCs. Developing countries were reluctant to accept apparent constraints on development for a problem not of their own making.[29]

In response to these concerns, the Montreal accord granted developing nations a 10-year grace period to meet the protocol's terms, and stipulated that industrial countries should provide funding and technology to help others make the transition. After the accord was finalized, however, it became increasingly clear that these provisions alone would likely be insufficient to convince many developing countries to join in. After some tough bargaining, an unprecedented global deal was struck in London in 1990: industrial countries agreed to reimburse developing countries for "all agreed incremental costs" of complying with the protocol—in other words, all additional costs above and beyond any they would expect to incur in the absence of the accord.[30]

Studies conducted by EPA indicated that these costs were relatively low—and paled in comparison with those that a damaged ozone layer would impose. The initial agreement called for the creation of a $240-million Interim Multilateral Fund. Key developing countries—including China and India—expressed satisfaction with the outcome, and announced their intention to join in the accord as a result. In 1992, governments agreed to make the Interim Multilateral Fund permanent. The fund has subsequently been replenished several times. As of mid-1999, industrial countries had contributed nearly $1 billion total, financing some 3,000 projects in 116 countries.[31]

Besides the "carrot" of funding, the Montreal Protocol also used the "stick" of possible trade restrictions. The negotiators were concerned about the potential for "CFC havens" to be created in countries that were not signatories—a development that could have seriously undermined the accord's effectiveness. To prevent this from happening, the protocol included provisions that forbade treaty members to trade in CFCs and products containing them with countries that have not joined the accord. Although effective in encouraging widespread participation, these provisions have become controversial in recent years, as international trade experts have questioned whether they would be permitted under the rules of the World Trade Organization. (See Chapter 7.)[32]

International trade and investment have had a significant impact on the move to phase out ozone-depleting substances in the developing world. Export-oriented developing countries have tended to be ahead of the phaseout curb, as selling goods in industrial countries requires keeping pace with developments there. China learned this lesson the hard way: its refrigerator exports declined by 58 percent between 1988 and 1991 as demand in industrial countries for refrigerators with CFCs plummeted. The government then moved aggressively to develop ozone-friendly refrigerators, and has said it will phase CFCs out faster than required under the protocol. Nonetheless, China is currently the world's largest CFC producer, accounting for some 30 percent of remaining CFC production, as well as more than 70 percent of halon use.[33]

The effort to reduce CFC use in developing countries has been aided by a tendency for multinational corporations to adopt the same ozone practices in their overseas operations as they use at home. In the Philippines, for instance, many foreign-owned electronics manufacturers had already eliminated most uses of ozone-depleting substances as solvents by 1995. Similarly, usage in Kenya fell by two thirds between 1989 and 1993, due at least in part to changes instituted by companies based in industrial countries.[34]

Although the jury is still out on the effectiveness of the North-South ozone partnership, preliminary signs are encouraging. The protocol required developing countries to freeze CFC consumption in mid-1999, and to phase it out altogether by 2010. As a group, developing countries are ahead of schedule. Their use of CFCs and halons increased by some 16 percent from 1986 to 1995, but the growth trend reversed in 1996, when usage fell by 6 percent. Botswana, Cameroon, Colombia, and Malta have already completely phased out CFCs, and Indonesia, the Philippines, Thailand, and Viet Nam have reportedly stopped using them except for servicing refrigerators and other essential uses.[35]

The ozone story offers some hope that the impasse over developing-country participation in the climate change treaty will yet be overcome. Foreign investment is in some cases already helping developing countries make the transition to a more climate-benign development path. Compact fluorescent light bulbs, for example, first produced in the United States, are increasingly manufactured in the developing world. In 1997, China made about 100 million of these energy-efficient bulbs—more than any other country. The funding and technology came in part through joint ventures with lighting firms based in Hong Kong, Japan, the Netherlands, and Taiwan. Compact fluorescents produced by joint ventures consistently outrank those of domestic companies in meeting performance standards such as efficiency and durability.[36]

Renewable energy components are also now being made in developing countries. India, for instance, has become a major manufacturer of advanced wind turbines with the help of technology obtained through joint ventures and licensing agreements with Danish, Dutch, and German firms. It has become the world's fifth largest wind power producer, with an installed capacity of nearly 1,000 megawatts.[37]

Despite the failure to broker a political deal on climate change, many developing countries are in fact already moving ahead with innovative policies and programs. Brazil recently eliminated oil subsidies, saving 4 million tons of carbon as well as more than $2 billion. Mexico has distributed 1.7 million efficient compact fluorescent light bulbs, offsetting 32,000 tons of carbon annually. And Costa Rica enacted a 15-percent carbon tax, with a third of its revenues channeled into tree planting projects by farmers.[38]

Perhaps the most encouraging news comes from China. Already the world's second largest emitter of carbon dioxide, projections suggest that China will surpass the United States and climb into first place within the next two decades. China's CO_2 emissions climbed steadily at a rate of some 4 percent a year over the last two decades, but in the last few years this trend turned around. In 1998, China's emissions dropped by 3.7 percent, despite robust economic growth of 7.2 percent. One important factor in the decline was a recent $14-billion cut in annual coal subsidies.[39]

FORGING A CLIMATE-FRIENDLY GLOBAL ECONOMY

As negotiators work to complete many details of the Kyoto Protocol by the end of 2000, it is becoming clear that climate change will be far more difficult to solve than ozone depletion was. CFCs were produced by a handful of major international companies that were able to switch with relative ease to even more profitable substitute chemicals. But carbon dioxide emissions are a ubiquitous byproduct of modern life. Sharply limiting them will require not only far-reaching technological transformations, but also lifestyle changes on the part of billions of people. Steady growth in purchases of sport utility vehicles as well as the size of homes in the United States are among the many factors driving CO_2 emissions on their upward course.[40]

Despite the many differences between the two problems, the Montreal accord does provide a key precedent for action on climate change. As Richard Benedick, chief U.S. negotiator for the Montreal Protocol, explains it, "by providing CFC producers with the certainty that their sales were destined to decline, the protocol unleashed the creative energies and considerable resources of the private sector in the search for solutions. The treaty at one stroke changed the market rules and thereby made research into substitutes economically worthwhile." If it is to succeed, the Kyoto Protocol will have to have a similar effect on the world's energy economy.[41]

The last few years have brought some encouraging signs that such a transformation is beginning to take hold. Markets for wind and solar power are booming at double-digit growth rates, while new industrial equipment and residential appliances are becoming steadily more efficient. In another promising development, most major auto makers have recently announced accelerated plans for the introduction of low-emission electric and fuel cell vehicles.[42]

And since 1997, even major fossil fuel companies, including British Petroleum, Royal Dutch Shell, and ARCO are beginning to figure Kyoto into their investment plans. British Petroleum has committed to building up its wind and solar energy businesses to at least $1 billion in annual sales over the next decade, and Shell has announced plans to invest $500 million over five years in renewable energy development. Both companies have also said they plan to cut their own greenhouse gas emissions by 10 percent, and have withdrawn from the Global Climate Coalition, an alliance of business groups that actively opposes the Kyoto Protocol. Mike Bowlin, the CEO of ARCO, went so far as to predict at a 1999 petroleum conference in Houston that "the last days of the age of oil" were near, and that oil companies should therefore broaden their energy investments.[43]

Despite these dramatic defections, major U.S. industry groups have continued efforts to sway public opinion against the climate change accord, in part by recruiting scientists to cast doubts on prevailing views about the seriousness of the problem. And despite studies indicating that the climate change treaty could lead to a net addition of nearly 800,000 new U.S. jobs, the AFL-CIO, which represents workers in industries such as mining and auto making, is maintaining its staunch opposition to the Kyoto Protocol, warning that it could have a "devastating effect on the U.S. economy and American workers."[44]

Although the stalemated climate politics of the past few years have led many observers to doubt whether the Kyoto Protocol will ever go into effect—and to question whether this may just be a problem so big that human society is not yet equipped to deal with it—the history of the Montreal process offers a sliver of optimism. As the ozone experience shows, scientific evidence can emerge unexpectedly and with dramatic effect, and politics can shift even more suddenly. The late 1990s already brought clear evidence that glaciers are melting worldwide, with scientists reporting a significant thinning in Greenland's ice sheet. Also, 1998's record temperatures appear to have precipitated a massive die-off in the world's ecologically rich coral reefs, with up to half now showing signs of temperature-induced "bleaching." The unprecedented storm damage of 1998 caused both untold human suffering and $92 billion in economic losses. Climate instability now threatens to rival financial instability for economic headlines in the early part of the new millennium.[45]

Even with such disturbing scientific developments, it may well be that it is the global economy rather than the global atmosphere that determines the outcome of the climate negotiations. At some point, key industries and governments may come to the same conclusion they reached with ozone depletion—that a major industrial transition is about to occur, and that the economic rewards will go to companies and countries that lead the way.

II

REFORMING GLOBAL GOVERNANCE

CHAPTER 7

TRADE WARS

In July 1999, the U.S. government decided to get tough with the countries of the European Union (EU), slapping 100-percent tariffs on $116.8 million worth of European imports, including fruit juices, mustard, pork, truffles, and Roquefort cheese. The European offense was its refusal to revoke a ban on the import of meat treated with growth hormones—a refusal that defied a World Trade Organization (WTO) ruling that the ban was an unfair barrier to U.S. and Canadian beef exports. The EU insists the ban is not an intentional trade barrier at all, but only a prudent response to public concern that eating hormone-treated beef might cause cancer and other health problems. As of December, the EU had refused to back down.[1]

The U.S. sanctions were greeted with widespread consternation in Europe, particularly in France, where a number of McDonald's restaurants were targeted for protests. In a symbolic tit-for-tat, the mayor of the French village of St. Pierre-de-Trivisy, in the heart of Roquefort cheese country, decided to retaliate by doubling the price of Coca-Cola sold at the town's campground and recreation center.[2]

This fierce transatlantic food fight is emblematic of a new kind of global trade conflict, in which health and environmental laws, rather than traditional trade issues such as tariffs, quotas, and the dumping of commodities like steel or wheat, are now at stake. The collision between the push for freer trade and an array of environmental laws implemented over the last several decades has spurred activists from around the world to sharply challenge the WTO's right to stand in judgment on national laws. Under pressure, policymakers are beginning to contemplate environmental modifications to the rules of world trade. But the campaign to "green" the WTO will likely be long and hard-fought.

FOOD FIGHTS

The accord that created the World Trade Organization included provisions that impose new restrictions on laws designed to protect human, animal, and plant health. Trade specialists had argued that legislators were passing disingenuous laws that lacked a scientific rationale, with the primary goal of keeping foreign products off their shelves. In order to prevent this kind of presumed interference with free trade, the Sanitary and Phytosanitary (SPS) Measures agreement encourages countries to harmonize a range of relevant standards at the international level. Food safety requirements are high on this list.[3]

Although the agreement permits countries to maintain national laws that are tougher than international standards, it places sizable legal hurdles in their way if they choose to do so. For instance, if an environmental law is challenged, the country defending the law must demonstrate that it is scientifically justified and based on risk assessment. Environmentalists and consumer groups argue that the new restrictions promote least-common-denominator policy-making—adoption of policies that are weak enough to be acceptable to the least environmentally protective member countries. The worry is that vested interests will exploit any scientific uncertainty surrounding a protective law (and in science, there are always uncertainties) as a reason to limit preventative environmental action.[4]

The need for such action is embodied in the precautionary principle—a basic tenet of international environmental law that is steadily gaining ground. The Rio Declaration on Environment and Development, for example, which was agreed to at the June 1992 Earth Summit, declares that: "Where there are threats of serious or irreversible damage, lack of full scientific certainty shall not be used as a reason for postponing cost-effective measures to prevent environmental degradation." The WTO's provisions, on the other hand, require that health and safety laws be based on scientific principles and not be maintained with insufficient scientific evidence. Although on the face of it these requirements sound reasonable enough, in practice countries often disagree about how much evidence is "sufficient" to justify preventative measures. The WTO shifts the burden of proof—in effect requiring that chemicals and other food additives be proved harmful before their use can be restricted. The problem with this approach is that extensive testing, sometimes over a period of years, is required to know if a substance has long-term cumulative effects that might cause cancer, damage to the immune system, or other serious ailments.[5]

As the Uruguay Round of trade negotiations was wrapping up in 1993, the European Community and the United States were already embroiled in a dispute over a European law that forbids the sale of meat produced using growth hormones—the dispute that is only now coming to a full boil. Since it went into effect in the late 1980s, the law has always applied equally to domestically raised and imported livestock, and has thus passed the WTO's bedrock test of nondiscrimination. The EU maintains that the ban is not an intentional trade barrier, but a prudent response to public concern that eating hormone-treated beef causes cancer and reproductive health problems. A number of studies suggest that the public concern may be justified. But the hormone-hooked U.S. livestock industry was threatened by the ban, as it blocks hundreds of millions of dollars worth of U.S. beef exports. The industry prevailed on the U.S. government to take up its cause at the WTO. The SPS agreement provided added ammunition for this long-standing U.S. campaign to use international trade rules to overturn the disputed European law.[6]

This effort led to a February 1998 WTO appeals panel ruling, which upheld an earlier dispute panel ruling that the European law violated WTO rules. The panelists' preliminary argument was that the law was based on inadequate risk assessment. They explicitly rejected the EU's defense that the import restriction was justified by the precautionary principle. Environmentalists were aghast at the decision. The U.S. consumer group Public Citizen charged that "through the dispute over hormone-treated beef, the WTO inappropriately inserted itself as a major arbiter of domestic health and safety policy. The WTO's beef hormone decision undermines countries' democratic prerogatives to safeguard their citizens' health and well-being."[7]

The beef hormone controversy is widely viewed as just a warm-up for a more serious trade controversy now brewing over genetically modified organisms (GMOs). Once again, the European Union and the United States are the primary antagonists. Prompted by public concern over the uncertain health and ecological effects of GMOs, the EU passed legislation in 1998 requiring all food products that contain genetically modified soybeans or corn to be labeled as such. Several other countries, including Australia, Japan, and

South Korea, are now following suit. A large share of food products made by U.S. companies—breads, salad oils, and ice cream, among them—now contain GMOs. Many European producers, in contrast, are steering clear of GMOs in the face of public concern. U.S. companies complain that the labeling requirements amount to trade barriers, and the U.S. and Canadian governments are now making this same point at the WTO and in other international forums.[8]

U.S. companies are also frustrated that the EU and much of the rest of the world have been slow to grant approval for the sowing of many varieties of genetically modified seeds, or the sale of crops grown from them. U.S. agribusiness has seen multimillion-dollar markets largely dry up as a result. U.S. exports of corn to Europe have virtually ceased, as genetically modified corn cannot be reliably separated from conventional varieties. The situation does not appear likely to change anytime soon. EU environment ministers agreed in June 1999 to a moratorium on new approvals while the EU's law governing the approval of GMOs is revised, a process that is likely to take two years or more.[9]

As in the beef hormone case, the U.S. government maintains that restrictions on GMOs violate WTO rules because hard scientific evidence of adverse health and ecological effects is lacking. The EU and various countries dispute this view, maintaining instead that labeling requirements and other policies are a prudent response to a new technology that has potentially large health and ecological effects that are still clouded by scientific uncertainties. These countries emphasize that the precautionary principle dictates proceeding with caution until more is known. Labeling, in particular, seems to many people around the world to be a reasonable response to consumer concerns. Many citizens view efforts to frustrate labeling laws as a profound threat to a consumer's "right to know."[10]

In February 1999, a proposed "biosafety" protocol to the

biological diversity convention was the first major victim of the burgeoning international trade war over GMOs. Negotiations under way for a few years had been aimed at putting in place a system of prior consent for the transport of genetically engineered seeds and products. The talks were scheduled to wrap up in Cartagena, Colombia, in February, but six major agricultural exporters—Argentina, Australia, Canada, Chile, the United States, and Uruguay—put a monkey wrench into these plans by blocking adoption of the accord. One of the main U.S. arguments against the protocol was a claim that its provisions ran counter to the rules of the WTO. As of December 1999, negotiators were still hoping to bridge the differences.[11]

Trade tensions over GMOs have also been simmering over the last few years within WTO committees. The issue is likely to receive more prominent treatment soon. Both the United States and Canada are pushing for it to be explicitly included on the agenda for future trade talks. And officials on both sides of the Atlantic are warning that the GMO issue could soon provoke a full-fledged transatlantic trade war.[12]

THE TUNA-DOLPHIN CHALLENGE

The aggressive U.S. stance against the EU's food safety laws is somewhat surprising, given that several of the nation's own environmental laws have also run afoul of world trade rules in recent years. In fact, widespread public concern about the environmental impact of the General Agreement on Tariffs and Trade (GATT)—the WTO's predecessor—was sparked in September 1991 by a GATT ruling against an embargo on Mexican tuna imposed under the U.S. Marine Mammal Protection Act (MMPA). This ruling raised widespread concerns that world trade rules will pose a barrier to effective action to protect global environmental resources.[13]

The United States imposed the trade restriction after

determining that Mexicans were fishing for tuna by a controversial method known as "setting nets on dolphins." For reasons that are not fully understood, dolphins and tuna tend to swim together in the eastern tropical Pacific Ocean. Fishers use the dolphins as markers for the tuna below. They set enormous purse-seine nets on schools of dolphins, bringing up tuna but also trapping dolphins in the process. In 1972, the original MMPA had sharply curtailed this practice for U.S. fishers by mandating tight restrictions on marine mammal mortality as a byproduct of commercial fishing. The results were dramatic: the number of dolphins ensnared by the U.S. tuna fleet fell from 368,600 in 1972 to under 10,000 in 1983.[14]

But dolphin advocates noticed a disturbing trend—foreign fishing boats began to account for a growing share of the tuna fleet in the eastern tropical Pacific, many of them formerly U.S. fishers who had simply reflagged their vessels in foreign ports. In order to prevent imports of foreign-caught, dolphin-deadly tuna from subverting the purpose of the law, the U.S. Congress amended the MMPA twice during the 1980s to impose restrictions on imported tuna comparable to those that applied to the U.S. fleet.[15]

In concluding that the tuna embargo breached current trade rules, the GATT panelists emphasized a key though controversial distinction between import restrictions aimed at the characteristic of products themselves and those keyed to production processes. They decreed that the U.S. law was illegal under GATT because the United States was rejecting the process by which the tuna were harvested rather than the tuna itself. Although GATT, and later the WTO, contains a specific provision that ostensibly protects the right of countries to pursue environmental protection policies that might otherwise contradict trade rules, the panelists ruled that this environmental exception pertains only to efforts by countries to protect the environment within their own borders. Because the Mexican tuna fishing took place outside of U.S. waters, the panelists viewed the embargo as tantamount to the United States foisting its environmental laws and values on the rest of the world. This point of view resonated with many people, particularly in the developing world, who looked to the rule-based GATT as a check on the U.S. tendency to wield its economic power unilaterally.[16]

But the decision exposed some glaring inconsistencies between the rules of the world trading system and emerging international environmental principles and practices. The trading system's aversion to process-related trade restrictions struck many people as particularly arbitrary, as environmental policy is moving increasingly toward focusing on the environmental impacts of products throughout their life-cycle—including production, distribution, use, and disposal. Gold or timber may be harmless or beneficial as products, for example, but enormously costly to human or environmental health in the ways they are processed, with gold extraction leaching cyanide into groundwater, and clearcutting reducing vast swaths of primary forests to wastelands. Reform of extraction and manufacturing processes is essential to making real environmental advances, yet trade rules put up a sizable hurdle to pursuing such efforts in a world economy that is becoming steadily more integrated.[17]

Also worrisome was the ruling's failure to acknowledge the right of countries to take action to protect the atmosphere, the oceans, and other parts of the global commons—a failure that raised questions about the legality under GATT of an array of environmental policies. What would become of policies aimed at reducing the use of harmful drift nets in fishing, protecting primary forests, or staving off ozone depletion or global warming? By the panel's reasoning, it seemed that even provisions of international environmental agreements designed to protect global resources could be ruled GATT-illegal. This clash between two different spheres of international law presented the world with a major legal challenge, as it is not always clear which agreement trumps the other in cases where two treaties are in conflict.[18]

It was thus somewhat ironic that the panelists further argued that the lack of an international agreement on dolphin protection practices in tuna fishing was one of the factors that made the U.S. import restriction suspect. Although coordinated international action to combat shared environmental threats is imperative, the process of reaching consensus can take years and even decades—time the world can ill afford in the face of accelerating, potentially irreversible ecological decline. At the time the U.S. import restrictions were imposed, nations had been trying for years to reach an agreement on dolphin-friendly fishing practices through the Inter-American Tropical Tuna Commission (IATTC). Twenty years after the United States first adopted dolphin-protection legislation, the IATTC finally reached an agreement in early 1991 requiring observers on boats and reduced dolphin mortality rates. But Mexico refused to participate at that time.[19]

In contrast to the notoriously slow pace of multilateral diplomacy, unilateral trade restrictions often generate quick and decisive results. In the case of the tuna-dolphin restrictions, several countries—including Ecuador, Panama, and Vanuatu—responded to the import restrictions by taking steps to limit dolphin mortality in tuna fishing. As "dolphin-safe" tuna fishing began to take off worldwide, reported dolphin mortality in the eastern tropical Pacific plummeted, falling from 133,000 in 1986 to less than 2,000 in 1998.[20]

Although unilateral trade restrictions are often derided as inimical to multilateral cooperation, they can in fact be an important tool for promoting stronger and more effective international accords. The tuna-dolphin controversy is a case in point. The 1991 decision was never formally adopt-

ed by the GATT Council, as Mexico and the United States
joined forces at GATT to prevent this from happening. (Neither country wanted to create a political firestorm in the
midst of efforts to get the North American Free Trade Agreement approved by the U.S. Congress.) A subsequent similar
ruling in 1994 on a case brought by the European Commission was also never formally adopted. But the controversy
over these cases did spur both countries to get serious about
reinvigorating negotiations through the IATTC, leading to a
binding agreement in February 1998 on an International
Dolphin Conservation Program among eight countries
whose vessels fish in the eastern Pacific Ocean.[21]

Signatories to this agreement, among them both Mexico
and the United States, agreed to a range of provisions,
including binding limits on dolphin mortality in the region;
mandatory observers on all vessels to monitor the agreement; reductions in the unintended "bycatch" of sharks,
billfish, and juvenile tuna in addition to sea turtles; and sustainable catch quotas for tuna. In return for these commitments, the United States pledged to lift the tuna embargo for
all participating countries. To honor this promise, the
administration worked with its allies in Congress to amend
the U.S. law to bring it into conformity with the International Dolphin Conservation Program.[22]

Some environmental groups bitterly criticized the
changes as a weakening of the law. The critics objected most
vociferously to the lifting of a prohibition on setting nets on
dolphins that was previously required to qualify for the U.S.
"dolphin-safe" label—and thus for access to the U.S. market.
Under the new agreement, the setting of nets on dolphins is
not prohibited so long as no dolphins are killed or seriously
injured in the process. Other environmental groups, including the usually uncompromising Greenpeace, supported the
International Dolphin Conservation Program and the associated changes in U.S. law. They maintain that the agreement

offers improved protection for dolphins and the broader
marine ecosystem. Many supporters of the law have also
expressed a preference for internationally accepted solutions
over unilaterally dictated policies.[23]

OF SHRIMP AND TURTLES

Despite the furor over the tuna-dolphin decision, in 1998 the
WTO ruled against a U.S. measure aimed at reducing unintended sea turtle mortality as a byproduct of shrimp trawling.
Sea turtles are both extremely endangered and highly mobile,
making international action to protect them a high priority.
The provisions of the U.S. law in question closed the lucrative U.S. shrimp market to countries that do not require their
shrimpers to use turtle excluder devices (TEDs)—simple but
highly effective pieces of equipment that prevent turtles from
getting ensnared in shrimp nets—or that do not have comparable policies in place. Spurred by the threat of U.S. trade
restrictions, 16 nations (13 in Latin America plus Indonesia,
Nigeria, and Thailand) have by now moved to require the use
of TEDs. India, Malaysia, and Pakistan chose a different tack,
however, deciding to launch a WTO challenge rather than
meeting the U.S. requirement. (Thailand joined them in this
effort as a matter of principle, even though it had adopted
TEDs.)[24]

Although the environmental effectiveness of the U.S. law
was clear, both the initial WTO dispute resolution panel and
a subsequent appeals panel concluded in 1998 that the measure violated WTO rules. The legal reasoning of the appeals
panel was an improvement over earlier rulings, as it
acknowledged that countries may in some circumstances be
justified in using trade measures to protect global resources.
But the panel nonetheless took issue with the way in which
the U.S. law had been implemented, arguing that it was
applied in an arbitrary manner that failed to treat countries

evenhandedly. The bottom line was that the U.S. law would have to be changed in order to comply with WTO rules. This outcome was particularly alarming, as the Uruguay Round of trade talks had strengthened the rules of dispute resolution proceedings to make rulings binding, and to provide for tougher trade retaliation in cases where countries are unwilling to change offending laws in order to adhere to panel findings.[25]

In response to the ruling, the U.S. government altered the way it was implementing the law without seeking any changes to the statute itself. The new guidelines provide for the import of specific shipments of shrimp that have been approved as turtle-safe even if the country as a whole has not met the certification requirements. The U.S. government also said it would step up its efforts to negotiate a multilateral accord on sea turtle protection with its Asian trading partners, although it is more likely that the Asian nations will press ahead on their own with a regional accord. It remains to be seen whether the U.S. response will satisfy the WTO, thus precluding the imposition of retaliatory trade sanctions.[26]

In any case, many U.S. environmentalists are unhappy with the government's response. Their primary concern is that the shipment-by-shipment method will be less effective in safeguarding turtles than the earlier blanket restriction, as it will not compel countries to mandate the use of TEDs when fishing for shrimp not destined for the U.S. market. A turtle might thus survive an encounter with a TED-equipped boat only to later fall prey to a TED-free vessel. Environmentalists also worry that the new policy may facilitate the entry of "laundered" shrimp into the United States. Several environmental groups filed suit against the government at the U.S. Court of International Trade, charging that the revised guidelines were inconsistent with provisions of the Endangered Species Act that stipulate adequate protection for sea turtles. In a preliminary ruling in April 1999, the court sided with the environmental groups, placing national law and international trade rules on a possible collision course.[27]

BEYOND SEATTLE

As opposition to the WTO continues to mount, many governments are beginning to acknowledge, rhetorically at least, that reforms are needed to make the world trading system environmentally sound. In a 1998 address commemorating the fiftieth anniversary of GATT, President Bill Clinton conceded: "We must do more to make sure that this new economy lifts up living standards around the world, and that spirited economic competition among nations never becomes a race to the bottom in environmental protections, consumer protections, and labor standards." And at the G8 summit meeting of industrial powers held in Cologne, Germany, in June 1999, world leaders agreed that environmental considerations should be taken into account in future WTO negotiations. But most governments have so far been vague about exactly how this should be done.[28]

Despite the lofty words, it remains far from clear that governments are now ready to amend existing WTO rules to buffer environmental laws from trade challenges. Among the priorities for reform are the following steps: clearly incorporating the precautionary principle into WTO rules, protecting consumers' right to know about the health and environmental impact of products they purchase by safeguarding labeling programs, recognizing the legitimacy of distinguishing among products based on how they were produced, ensuring the right of countries to use trade measures to protect the global commons, and providing deference to multinational environmental agreements in cases where they conflict with WTO rules. The European Union has voiced general support for many of these ideas, but the United States appears lukewarm about writing any new environmental guarantees into the WTO. Ongoing controversies over beef hormones and GMOs undoubtedly color the U.S. view.[29]

Other issues also cry out for attention. Developing countries are particularly concerned about that the Agreement on Trade-Related Intellectual Property Rights (TRIPS), which came out of the Uruguay Round. This agreement requires WTO members to adopt intellectual property rights systems largely patterned on the industrial-country model, or else to subject themselves to retaliatory trade sanctions. Developing countries argue that the TRIPS agreement contradicts provisions of the U.N. Convention on Biological Diversity that protect indigenous knowledge. At particular issue is a provision that requires countries to recognize the intellectual property of commercial plant breeders, while failing to provide for any remuneration to farmers who have gradually improved plant varieties over the course of centuries. Governments began a review of the TRIPS agreement in 1999. Many developing countries and nongovernmental organizations are pushing for the agreement to be substantially overhauled.[30]

On the more positive side, the WTO could conceivably be enlisted in an effort to reduce environmentally harmful subsidies. World trade rules have long discouraged subsidies, as they distort the economic playing field. The United States and six other nations have suggested building on this tradition by making the elimination of fishing subsidies an objective for future trade talks. These subsidies, which add up to some $14–20 billion annually, help propel overcapacity in the world's fishing fleet, which is itself a powerful driving force behind today's depleted fisheries. But the European Union and Japan, both of which are major subsidizers of their fishing fleets, have so far reacted cautiously to the initiative.[31]

Other environmentally harmful payouts could also be tackled at the WTO. The United States and other agricultural exporters are urging greater attention to the adverse environmental impacts of agricultural export subsidies. But the European Union and Japan are unenthusiastic, as they both face heavy domestic political pressure to maintain agricultural subsidies. Energy subsidies would be another logical candidate. They currently drain national treasuries to the tune of $100 billion annually, while also imposing heavy environmental costs. Subsidies to the timber industry, such as selling timber from state lands to private companies at cut-rate prices, could also potentially be taken on at the WTO.[32]

Procedural questions also urgently need to be addressed. The recent spate of environmentally related trade disputes has opened the WTO to intense scrutiny, with critics charging that its secretive ways pose a basic threat to democracy. Many important documents are unavailable to the public, and most WTO committees, as well as all dispute resolution proceedings, are conducted in closed sessions dominated by trade rather than environmental experts.[33]

Even *The Economist*, which normally pushes a free-trade agenda with nearly religious zeal, acknowledges that "the four-year old WTO is at a crossroads. It has become a quasi-judicial body, an embryo world government....Yet it is now being asked to arbitrate on matters that are intensely political. It lacks the legitimacy to do so." Indian activist Vandana Shiva makes essentially the same point, although she carries it a step further: "The WTO is basically the first constitution based on the rules of trade and the rules of commerce. Every other constitution has been based on the sovereignty of people and countries. Every constitution has protected life above profits. But WTO protects profits above the right to life of humans and other species."[34]

The unprecedented protests by tens of thousands of citi-

zen activists in Seattle in November 1999 were a powerful
wake-up call about the deep-seated public opposition to
international governance based on such a narrowly economic conception of the global interest. Far-reaching reforms are
needed if trade agreements are to garner the political support
they will need in order to be sustained in the new century.[35]

CHAPTER 8

GREENING THE FINANCIAL ARCHITECTURE

During the 1990s, money became increasingly mobile due to a range of factors, including the takeoff in computerized trading as well as the deregulation of international capital markets. International investment surged in response, particularly into the newly established stock markets of the developing world.[1]

Private capital flows into developing countries and the former Eastern bloc increased from $53 billion at the beginning of the 1990s to an all-time high of $302 billion in 1997. Although just 44 percent of the capital moving into the developing world in 1990 was from private sources, by 1997 the figure reached 88 percent. (See Figure 8–1.) Large parts of Asia and Latin America were suddenly transformed in the minds of international investors from poor, "developing countries" into glistening "emerging markets." At the same time, new financial instruments such as hedge funds and derivatives created an explosion of foreign exchange trading, with $1.5 trillion changing hands every day.[2]

But in 1997 the bubble burst. Thailand was the first eco-

FIGURE 8–1

Private Capital Flows to Developing Countries, 1970–98

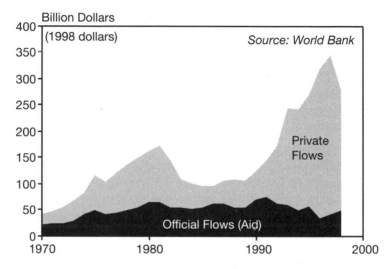

nomic domino to fall, when it was forced to devalue its currency sharply after it came under sustained speculative attack. The crisis soon spread to Indonesia, Malaysia, the Philippines, and South Korea. International investors lost their nerve and raced for the exits. Some $20 billion in 1997, and $30 billion in 1998, flowed out of the Asian countries in crisis. Banks failed and stock markets collapsed, sending the economies of the region into a tailspin.[3]

And the crisis did not stop at the continent's edge. Shaken by the Asian experience, investors began to pull money out of emerging markets everywhere. Russia's currency and stock market went into a free fall in late August 1998, forcing the country to default on $40 billion in international loans. Brazil appeared to be the next domino waiting to fall, prompting the International Monetary Fund (IMF) to step in with a $42-billion bailout plan.[4]

The social and environmental fallout from the crisis

was severe. In battered Asia, tens of millions of people fell into poverty as their jobs disappeared, and as many as a million children were pulled out of school, with some of them pushed into prostitution by their desperate families. Growing poverty attributable to the crisis also had environmental costs, such as a surge in Indonesia in the poaching of endangered monkeys, tigers, and other species as destitute people sought quick cash. And governments and businesses desperate to stave off bankruptcy slashed environmental spending.[5]

As of late 1999, the global economic crisis appeared to be in at least temporary remission. International capital had begun to return to most of the countries affected by the crisis, and economic growth rates were again headed upward, although poverty rates remained stubbornly high. But many commentators warn against a false sense of complacency. They predict that the next jolt is not far off, as the globalization of international finance has outpaced the ability of governments and international institutions to manage the system effectively.[6]

If there is a silver lining to the crisis, it lies in the wake-up call sounded about the risks of rapid globalization, and the launching of a critical dialogue about how to reform what is being called the "international financial architecture" to meet the demands of the twenty-first century. As these discussions proceed, the need to protect the natural resource base that underpins the global economy merits a prominent place on the agenda.[7]

ENVIRONMENTAL ADJUSTMENT

The International Monetary Fund and the World Bank are both key pillars of the current international financial architecture. The missions of these two institutions, which were created in response to the Great Depression that helped precipitate World War II, are now being called into question. The IMF came under close scrutiny in the wake of the Asian economic crisis. The Fund's high-profile role as a conduit for multibillion-dollar bailout packages for the crisis-stricken countries was a clear demonstration of its formidable powers. But this also stirred controversy, as prominent economists took issue with the wisdom of the institution's financial advice and the secrecy shrouding its operations.[8]

Although the World Bank maintained a lower profile during the economic crisis of the late 1990s, it has also been active in channeling funds into the crisis-ridden countries, often in close cooperation with the IMF. Over the last few years, the Bank has boosted both its total lending and the share of its funds that are spent on cash infusions for "structural adjustment." The Bank's total lending reached $29 billion in 1999, up from just $19 billion two years earlier. More than half of this was for structural adjustment, compared with only 27 percent in 1997. Under conventional structural adjustment loans as well as the crisis-generated bailout packages, countries receiving funds agree to implement a long and specific list of policy changes intended to restore them to economic health and thus to creditworthiness. Privatization, price and exchange rate stability, and trade liberalization are among the policies commonly recommended.[9]

But the World Bank and the IMF pay insufficient heed to the profound effects of these policies on the ecological health and social fabric of recipient countries. One important component of most adjustment loans is policies aimed at raising exports in order to generate foreign exchange with which to pay back debts. Yet the pressure to export can lead countries to liquidate natural assets such as fisheries and forests, thereby undermining longer-term economic prospects. Intensive export-oriented agriculture is also often promoted, sometimes at the expense of small farmers and indigenous peoples. At the same time that structural adjustment loans promote exports of environmentally sensitive commodities, they also often require countries to make Draconian cuts in government spending. The budgets of already overburdened environment and natural resource management ministries are rarely spared.[10]

All these effects are evident in the recent bailout packages. In Indonesia, the IMF encouraged more palm oil production as part of its broader strategy for pulling that country out of its economic crisis, mandating that Indonesia remove restrictions on foreign investment in this sector. Yet rapid growth in palm oil exports has been a major contributor to the decimation of Indonesia's biologically rich tropical forests in recent years, raising questions about the wisdom of pushing such exports further still. (See Chapter 4.)[11]

Spending on environmental protection has declined markedly in the crisis-ridden countries, including Indonesia, the Philippines, South Korea, and Thailand. In Russia, the budget for protected areas was recently cut by 40 percent, a move that nongovernmental organizations (NGOs) in the region attribute to pressure from the IMF. And as part of a recent pact with the IMF, Brazil agreed to major cutbacks in environmental and social spending. A key international program aimed at protecting the Amazonian rainforest from destruction by ranchers, loggers, farmers, and miners is one of the programs that faced the chopping block, although a public outcry succeeded in saving the initiative from complete collapse. The timing of these reductions was particularly poor in light of the high rates of deforestation in the Brazilian Amazon over the last few years.[12]

Although structural adjustment programs often lead to environmental harm, they have also been used in a few cases to promote environmentally beneficial policy changes. In 1996 and 1997, the IMF suspended loans to Cambodia after government officials awarded logging concessions to foreign firms that threatened to open up the country's entire remaining forest area to exploitation—while funneling tens of millions of dollars into the bank accounts of corrupt officials. And despite its worrisome provisions for stepped-up natural resource exports, the recent Indonesian bailout plan also included several provisions intended to benefit forest management in the country.[13]

As part of an assault on the country's tradition of "crony capitalism," the Indonesian bailout plan required a number of reforms to the country's corruption-laden forestry sector, including tighter control over a government reforestation fund, the revenues of which had often just lined the pockets of President Suharto's political allies. The bailout package also included several measures aimed specifically at protecting the country's forests, such as reducing land conversion targets to environmentally sustainable levels, instituting an auctioning system for handing out concessions, and imposing new taxes on timber sales. Although these reforms were a step in the right direction, their ultimate effect on deforestation rates in the country remains to be seen.[14]

Besides using their influence to discourage unsustainable levels of natural resource exploitation, the IMF and the World Bank are well placed to promote environmentally beneficial fiscal reforms, such as cuts in environmentally harmful subsidies or the imposition of pollution taxes. They could also promote improvements in environmental accounting, such as incorporating the depletion of natural resources into national income figures. As things now stand, the destruction of natural assets such as forests, fisheries, and minerals is not typically included in national income figures, which means that policymakers are working from an incomplete set of books. And the IMF could include environmental issues in its mandate to monitor the economic prospects of its member countries, in part by tracking environmental spending levels and structural adjustment–mandated legislative changes that affect the environment.[15]

Despite the clear links between economic and environmental health, the IMF has long resisted the idea that environmental issues have much to do with its mission. When the organization was first created, its primary role was to help tide countries over when they faced short-term liquidity problems rather than to help them meet longer-term development goals. But the abandonment of fixed exchange rates in the 1970s deprived the IMF of much of its original mandate. Since then it has become increasingly involved with issues of longer-term development, such as its prominent role in brokering debt restructuring deals in the 1980s. The IMF now accepts that issues such as fighting corruption and alleviating poverty intersect with its mission. It is difficult to see why environmental protection should be any different.[16]

On paper, the development-oriented World Bank has been far more open than the IMF to the idea that environmental concerns should be integrated into its structural adjustment lending. The Bank's policy that governs adjustment lending notes that the environmental impact of loans should be fully considered as they are prepared, with a view toward promoting possible synergies and avoiding environmentally harmful results. But an internal review by the Bank of more than 50 recent loans found that few paid much heed to environmental and social matters. Whereas a 1993 Bank report found that some 60 percent of adjustment loans included environmental goals, the recent study concluded that this share had plummeted to less than 20 percent. An added problem is the fact that the Bank's policy on environmental impact assessment does not cover broad-based structural adjustment lending, although it is supposed to apply to adjustment loans aimed at specific sectors, such as agriculture and energy. IMF loans are also not subject to environmental impact assessment.[17]

LEVERAGING CHANGE

Despite the World Bank's growing role in adjustment lending, project lending remains central to its activities. The
Bank has traditionally made loans only to governments, but
in the last few years it has increasingly emphasized support-
ing the private sector. It has done this both by using its own
funds to guarantee private-sector projects and by bolstering
the operations of two affiliated agencies, the International
Finance Corporation (IFC) and the Multilateral Investment
Guarantee Agency (MIGA). The IFC directly finances private enterprises, while MIGA insures corporations against
political risks, such as expropriation, civil disturbance, and
breach of contract. At last count, in 1995, the World Bank
estimated that some 10 percent of all private-sector investment in the developing world was supported at least indi-
rectly by its various private-sector programs.[18]

Both IFC and MIGA are involved in many large investment projects with heavy ecological impacts. Nearly half of
IFC's committed portfolio of $12.9 billion in 1999 was
invested in sectors with large environmental impacts, such as
automobile manufacturing, chemicals, construction, infrastructure, and mining. In recent years, the IFC has helped
finance a chemical factory in Venezuela, a gold mine in Burk-
ina Faso, and a cement factory in China, to cite just a few
examples.[19]

After more than a decade of pressure from NGOs and
determined efforts by committed insiders, the World Bank
now has an extensive set of environmental and social policies, which among other things cover environmental impact
assessments of projects, forestry lending, involuntary resettlement, protection of wilderness areas, the rights of indigenous peoples, and pest management. The IFC and MIGA
both recently issued their own parallel policies, and the
World Bank published an updated Pollution Prevention and

Abatement Handbook, which provides detailed pollution reduction guidelines for nearly 40 industries. The importance of the World Bank's standards is magnified by the fact that these are often looked to by private investors as the prevailing international norm.[20]

In theory, all Bank agencies are supposed to bound by their policies, although the Bank admits it has a tarnished history in following its own rules. In recent years, the organization has taken steps to improve compliance, including creating a special unit charged with overseeing implementation of 10 "safeguards" policies, which include most environmental and social requirements.[21]

But controversies continue to swirl around ongoing and proposed projects, raising questions about the seriousness of the Bank's newly professed commitment to enforcing its policies. In June 1999, the Board of Directors approved a controversial $160-million loan to China for the resettlement of some 58,000 poor farmers to a sparsely populated part of Tibet, over the objections of both the German and U.S. governments. Critics charge that the planned resettlement is mainly aimed at helping China exert political control in the region, and that it will cause substantial environmental and social disruption, in violation of the Bank's own policies.[22]

The Bank is also considering a package of support for a $3.5-billion project to build an oil pipeline through untouched rainforest in Cameroon to oilfields in southern Chad. Environmentalists charge that the project amounts to "corporate welfare"—at large environmental and social cost. They also argue that the proposed project would not comply with several Bank policies, including those on environmental impact assessment and protection of the rights of indigenous peoples. The project sponsors initially included Elf Aquitaine, Royal Dutch Shell, and Exxon, although Elf and Shell withdrew in late 1999. It is not clear whether their

decisions were influenced by the opposition of environmentalists or by economic factors alone. As of late 1999, the
Bank's Board of Directors had not yet decided whether to
approve the project.[23]

As World Bank environmental and social standards were
strengthened over the last decade, private investors turned
increasingly to bilateral export financing agencies to find
support for projects that no longer passed muster at the
Bank. Export credit support climbed from $24 billion in
1988 to $105 billion in 1996. All told, bilateral export pro-
motion in the form of loans and investment insurance now
underwrites more than 10 percent of all world trade. Bilateral export promotion often supports environmentally disruptive projects, including mines, pipelines, power plants,
and hydroelectric dams.[24]

The U.S. government has had environmental policies in
place for several years at two of its export promotion agencies, the U.S. Overseas Private Investment Corporation
(OPIC) and the U.S. Export-Import Bank. Both agencies
have strengthened their policies in recent years to, among
other things, require the agencies to track and report on
greenhouse gas emissions from projects they support, and to
prohibit support for logging in primary tropical forests.[25]

Although these policies are important steps in the right
direction, the last few years have provided ample evidence of
their shortcomings. A 1999 report by the Washington-based
Institute for Policy Studies and Friends of the Earth concluded that OPIC and the Export-Import Bank between
them underwrote some $23 billion in financing for oil, gas,
and coal projects between 1992 and 1998. Over their life-
times, these projects will release some 29 billion tons of carbon dioxide, more than total global emissions in 1998.[26]

And in June 1999, OPIC's Board of Directors approved a
controversial $200-million package of loan guarantees to
Enron, Shell, and local Bolivian partners for the construc-

tion of a 630-kilometer gas pipeline from eastern Bolivia to a power plant in Cuiabá, Brazil. The project will cut through some of Latin America's most important natural areas, including Bolivia's Chiquitano Forest and the headwaters of the Pantanal, the world's largest wetland. Environmentalists charge that the loan was made in blatant contradiction of OPIC's prohibition on financing infrastructure and extractive projects in primary tropical forests. OPIC counters that the affected forest is not "primary," as some isolated logging has already occurred in the area. It also touts the environmental benefits of natural gas, which the agency argues will reduce demand for fuelwood and diesel fuel.[27]

Even with the best of policies, in a global economy tough national standards can easily be undermined by laggards abroad. The United States learned this lesson the hard way a few years back when its Export-Import Bank refused on environmental grounds to extend credits to companies such as the heavy equipment manufacturer Caterpillar that wanted to participate in China's Three Gorges Dam project. The dam is expected to flood more than 60,000 hectares of land and 160 towns, forcing the resettlement of some 1.3 million people. But the Bank's counterparts in Canada, France, Germany, Japan, and Switzerland stepped into the breach created by the U.S. decision. Stung by the experience, the United States is working to persuade other donor countries to develop environmental guidelines for their export finance agencies. Several countries are now developing such standards, including Canada, Japan, Norway, and the United Kingdom. Negotiations are also under way to create common environmental standards for the export finance agencies of the major industrial countries. NGOs are pushing for them to be set at a high level.[28]

THE GREENING OF WALL STREET?

Although strengthened environmental standards at the world's export credit agencies are desperately needed, the risk remains that private capital markets will be tapped for environmentally damaging projects. In the Three Gorges case, a number of prominent investment banks—including Lehman Brothers, Morgan Stanley, and Salomon Smith Barney—have sponsored bond offerings over the last few years to help the Chinese government raise funds for the dam. Although convincing private financiers to pay attention to the environment is substantially more difficult than lobbying public institutions such as the World Bank and export credit agencies, environmental activists and others are pushing private lenders and investors to pay attention to the environmental consequences of their loans.[29]

Commercial banks require exhaustive studies of possible risks before making loans, a process known as "due diligence." Increasingly, banks are viewing environmental issues as an important consideration in this process. They have a diverse range of concerns. Banks worry, for example, that a hazardous waste dump will be discovered on a property they lent money for, and that they will be held liable, as has happened in recent U.S. court cases. They also fear that violations of environmental laws will lead to large financial penalties that will undermine a borrower's creditworthiness. In the most extreme case, a project might be stopped altogether in the face of opposition from local citizens and environmental groups.[30]

"International commercial banks, whether they intend to be or not, are frequently very effective enforcers of local and international environmental requirements," maintains Bradford Gentry of Yale University. "The level of scrutiny given to these issues by banks is often well above that of local environmental enforcement." Nonetheless, a 1997 study by the National Wildlife Federation found substantial room for improvement: fewer than half of 51 financial institutions from 13 countries on four continents routinely conduct environmental due diligence on transactions other than those secured with real estate.[31]

In an effort to better this record, the U.N. Environment Programme (UNEP) launched an effort in 1992 to encourage major banks around the world to incorporate environmental considerations into their lending programs. So far, 162 banks from 43 countries have signed the initiative's Statement by Banks on the Environment and Sustainable Development. The signatories underscore their expectation that borrowers must comply with "all applicable local, national, and international environmental regulations." They also pledge to update their accounting procedures to reflect environmental risks, such as the potential for chemical accidents or hidden hazardous waste dumps, and to develop banking products and services that promote environmental protection.[32]

Although laudable in its goals, the UNEP statement requires few specific commitments. In fact, several signatories were involved with the recent Chinese bond offerings that activists charge are funneling money into the Three Gorges project. In order to avoid such gaps between rhetoric and reality, the U.K.-based Green Alliance suggests strengthening the UNEP initiative by transforming the statement into a document whose expected standards of performance are clear enough to be audited.[33]

Stock market investors are also slowly beginning to show more interest in environmental questions. It used to be assumed that it was costly for companies to be good environmental stewards. But this view is giving way to new evidence that companies with strong environmental management structures may in fact perform better financially, on average, than companies that are plagued by large environmental liabilities such as the threat of paying costly fines.[34]

A 1995 report by the Investor Responsibility Research Center compared the stock market performance of the companies in the Standard & Poors index, a group of 500 representative stocks. They sorted the firms into "high-" and "low-" polluting companies. Overall, the study found no penalty for investing in "green" portfolios, and concluded that in some cases low-pollution portfolios actually demonstrated superior performance. A November 1996 study by the consulting firm ICF Kaiser was more bullish still. Its survey of more than 300 Standard & Poors companies revealed that adopting proactive environmental policies had a "significant and favorable impact" on a firm's value in the marketplace, as it reduced the perceived risk of investing in the company, and thus its cost of borrowing money.[35]

As studies like these accumulate, environmentally screened investment funds will likely grow in popularity. A frontrunner is the Global Environment Fund, a private, Washington-based investment fund manager founded in 1989. In part with the help of loan guarantee agreements with OPIC, the group has raised more than $500 million in investment capital from institutional investors for five investment funds, including two Global Environment Emerging Markets Funds, which between them now have holdings in some 10 countries in Africa, Asia, Eastern Europe, and Latin America. The principal focus of these funds is environmentally related infrastructure, including renewable energy projects and water and sewage treatment plants. A number of other green investment funds have been established over the last decade, including the Storebrand Scudder Environmental Value Fund in Luxembourg, the Sustainable Performance Group in Switzerland, and the Green Century Balanced Fund in the United States.[36]

In many cases, green investing is part of a broader socially responsible financial strategy. In the United States, socially responsible investing is a growth industry. Investment funds screened according to some social criteria surged from just $162 billion in 1995 to nearly $1.5 trillion in 1999—some 9 percent of all managed funds. The performance of these funds, like all investment funds, varies widely. Overall, however, the returns have been competitive. For instance, the Domini 400 Social, an index of socially screened firms, outperformed the Standard & Poors 500 over much of the 1990s.[37]

Even the venerable Dow Jones index is now getting in on the act. Dow Jones Indexes and the Switzerland-based SAM Sustainability Group launched a new Dow Jones Sustainability Group Index in September 1999 composed of more than 200 "sustainability-driven" companies, which the group defines as those that seek to "achieve their business goals by integrating economic, environmental, and social growth opportunities into their business strategies." The companies included in the new index represent 68 industries from 22 countries, with a collective total market capitalization of more than $4 trillion. The index is currently dominated by European companies in such sectors as automobiles, paper products, food, banks, insurance, and waste management. If it had existed over the past five years, the index would have outperformed conventional indices by more than 5 percent. Financial institutions in Germany, the Netherlands, and Switzerland are already creating investment funds based on the Sustainability Group Index.[38]

In addition to socially screened funds, another $922 billion—more than 5 percent of managed funds—is controlled by activist shareholders who try to influence the policies of the companies in which they own shares by participating in shareholder resolutions and proxy votes, among other techniques. In 1999, concerned investors introduced 54 shareholder resolutions related to environmental issues, and some of them have already borne fruit. In one particularly successful case, Home Depot announced a commitment to purchasing certified timber just three months after 12 percent of its shareholders asked the company to stop selling wood from old-growth forests. The shareholder resolution was one part of a broad campaign by activists to convince the company to take this step.[39]

Green investors need better information about corporate environmental performance if financial markets are to reflect environmental risks adequately. The last several years have seen an explosion of interest in environmental reporting, but existing efforts have been poorly coordinated, leading to a proliferation of "non-standardized information reported in non-uniform formats," according to the Coalition for Environmentally Responsible Economies (CERES). In an effort to address this deficiency, CERES launched a Global Reporting Initiative in 1997, in which corporations, NGOs, professional accounting firms, and UNEP are working together to produce a global set of guidelines for corporate sustainability reporting. The goal of the initiative is to elevate environmental reporting to the same plane as financial reporting, making it standard business practice worldwide.[40]

Reforms in financial accounting rules and regulations could help accomplish this. Companies operating in the United States are required to disclose large environmental liabilities, such as hazardous waste sites, on the forms they file with the Securities and Exchange Commission (SEC). But the information varies widely in quality, with many companies submitting vague boilerplate language or no data at all. The reports are particularly sketchy about performance overseas: a 1997 survey by the Investor Responsibility Research Center found that 73 out of 97 companies with foreign operations failed to include information about their environmental track records abroad in publicly available documents such as SEC submissions. The Corporate Sunshine Group, a U.S. alliance of investors, environmental organizations, community groups, and labor unions, is pushing the SEC to do a better job enforcing existing environmental and social disclosure rules, to broaden and deepen these requirements, and to enter into an information-sharing agreement with the U.S. Environmental Protection Agency on companies' environmental liabilities.[41]

As they write rules governing newly established stock markets, developing countries have an opportunity to get these policies right from the beginning. Thailand, for one, requires companies listed on the Stock Exchange of Thailand (SET) to undergo an environmental audit that includes an environmental impact assessment as well as a site visit. The SET also relaxes the stock-exchange-listing requirements for companies seeking finance for selected environmental control and prevention projects.[42]

International discussions on a new international financial architecture have focused heavily on the need for more transparency of financial data, so that investors can assess risk better, thereby avoiding financial meltdowns. But there has been little if any acknowledgment in these discussions that environmental risks will also pose a serious threat to international stability in the new millennium. Building an environmentally sustainable global economy will require rewriting the rules of international finance to account for environmental vulnerabilities.[43]

CHAPTER 9

STRENGTHENING GLOBAL ENVIRONMENTAL GOVERNANCE

Forging an environmentally sustainable society is about more than economics, and farsighted economics is about more than reducing restrictions on the movement of goods and money. Creating a global society fit for the twenty-first century will require not only reform of economic institutions, but also the creation of an international environmental infrastructure that can act as an ecological counterweight to today's growing economic powerhouses.

A good place to start is with the hundreds of agreements, declarations, action plans, and international treaties on the environment that now exist. Environmental treaties alone number more than 230; agreement on three fourths of them has been reached since the first U.N. conference on the environment was held in Stockholm in 1972. (See Figure 9–1.) These accords cover atmospheric pollution, ocean despoliation, endangered species, hazardous waste trade, and Antarctica, among other issues.[1]

The vast majority of environmental agreements are bilateral or regional in scope, involving, for instance, the management of river systems, air corridors, or migratory bird species. However, a minority of environmental issues—including the atmosphere, international waterways, and biological diversity—are truly global. The last few decades have seen steady progress toward developing international rules governing these "global commons." (See Table 9–1.)[2]

Judging from the number of treaties, environmental diplomacy appears to have been a spectacular success. And many of these accords have in fact had important results. Among other achievements, air pollution in Europe has been reduced dramatically as a result of the 1979 treaty on transboundary air pollution; global chlorofluorocarbon (CFC) production has dropped 87 percent from its peak in 1988 as a result of the 1987 Montreal Protocol on ozone depletion; the killing of elephants plummeted in Africa following a 1990 ban on commercial trade in ivory under the Convention on International Trade in Endangered Species of

FIGURE 9–1

International Environmental Treaties, 1920–98

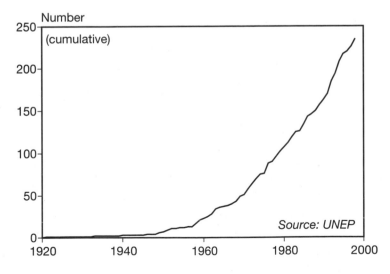

Source: UNEP

TABLE 9–1

Selected International Environmental Agreements

International Whaling Convention, 1946 (40 Parties)

Protects whales from excessive harvesting. Amended in 1992 to ban all commercial whaling of large whales from 1986 on.

Antarctic Treaty, 1959 (44 parties)

Subjects continent to joint management by original 12 parties and others who conduct scientific research there "in the interests of all mankind." Protects Antarctica from military activities, nuclear tests, and radioactive waste imports. Protocol signed in 1991 bans mining exploration and development for 50 years, protects wildlife, regulates waste disposal and marine pollution, and provides for scientific monitoring of the continent.

Convention on International Trade in Endangered Species of Wild Fauna and Flora (CITES), 1973 (146 parties)

Restricts trade in species that are either threatened with extinction or may become endangered if their trade is not regulated. Ban passed in 1990 prohibited trade in ivory.

U.N. Convention on the Law of the Sea, 1982 (132 parties)

An umbrella agreement that provides a broad framework for ocean development and establishes 200-mile Exclusive Economic Zones (EEZs). Includes provisions on conservation of living resources of the oceans, maintenance and restoration of populations of species, and protection of the sea from pollution.

Montreal Protocol on Substances That Deplete the Ozone Layer, 1987 (172 parties)

With amendments, requires phaseout of CFCs in industrial countries by 1996 and in developing countries by 2010. Also restricts the use of several other ozone-depleting substances.

Basel Convention, 1989 (133 parties)

Controls the transboundary movement of hazardous wastes. Amended in 1995 to completely ban exports to developing countries for final disposal and recovery operations.

TABLE 9–1 *(continued)*

U.N. Framework Convention on Climate Change, 1992 (180 parties)

Includes target for industrial countries of stabilizing carbon dioxide emissions at 1990 levels by 2000. Requires developing countries to undertake emissions inventories and other studies. 1997 Kyoto Protocol mandates stronger emissions reductions of 6–8 percent by 2008–12 for industrial countries, in order to meet the treaty's goal of stabilizing the overall concentration of carbon dioxide in the atmosphere.

Convention on Biological Diversity, 1992 (176 parties)

Establishes broad framework for the conservation of biological diversity, the sustainable use of its components, and the fair and equitable sharing of the benefits arising out of the use of genetic resources. Recognizes national sovereignty over biological resources. Jakarta Mandate of 1995 addresses the protection of marine and coastal diversity.

Convention on Desertification, 1995 (159 parties)

Combats desertification by promoting "bottom up" strategies focused on sustainable management of land and water. Supplies framework for local projects, encourages national action programs, establishes mechanism to coordinate funds, and encourages trust funds.

Agreement Relating to the Conservation and Management of Straddling Fish Stocks and Highly Migratory Fish Stocks, 1995 (24 parties)

Reinforces Law of the Sea convention to address the overexploitation of high seas fisheries by prescribing a precautionary approach to fishery management. Grants parties the right to board and inspect vessels of other parties, and obligates parties to collect and share data as well as to minimize bycatch of nontarget marine species. (Not yet in force—requires 30 ratifications.)

Rotterdam Convention on the Prior Informed Consent Procedure for Certain Hazardous Chemicals and Pesticides in International Trade, 1998 (2 parties)

Restricts the international export of 27 harmful pesticides and industrial chemicals that have been banned or severely restricted domestically, unless the importing country agrees to accept them. Creates a mechanism for information exchange. (Not yet in force—applied on a voluntary basis until 50 ratifications are obtained.)

SOURCE: See endnote 2.

Wild Flora and Fauna (CITES); the annual whale take
declined from more than 66,000 in 1961 to some 1,500
today as a result of agreements forged by the International
Whaling Commission; and mining exploration and development have been forbidden in Antarctica for 50 years under a
1991 accord.[3]

Yet even as the number of treaties climbs, the condition
of the biosphere continues to deteriorate. As noted in earlier chapters, carbon dioxide levels in the atmosphere have
reached record highs, scientists are warning that we are in
the midst of a period of mass extinction of species, the
world's major fisheries are depleted, and water shortages
loom worldwide. The notoriously slow pace of international diplomacy needs to be reconciled with the growing
urgency of protecting the planet's life-support systems.

Environmental treaties have so far mostly failed to turn
around today's alarming environmental trends because the
governments that created them have generally permitted
only vague commitments and lax enforcement. Governments have also for the most part failed to provide sufficient
funds to implement treaties, particularly in the developing
world. Ironically, environmentalists need to take a page from
the World Trade Organization (WTO) and push for international environmental commitments that are as specific and
enforceable as trade accords have become.

MAKING ENVIRONMENTAL TREATIES WORK

Reaching agreement on a treaty is only the first step. The real
work involves updating it continuously in light of new scientific information or changing political circumstances, and
ensuring that paper commitments are translated into real
policy changes in countries around the world.

Environmental treaties rely heavily on transparency as an

implementation tool. They generally require detailed reporting of actions taken at the national level to put agreements into practice. If this information is made freely available, then other countries as well as nongovernmental organizations (NGOs) can use it to shame countries into compliance.[4]

But governments often fail to provide secretariats with accurate, complete, and timely information. Only 51 percent of the parties to the Convention on Biological Diversity had submitted the required reports as of March 1998, for instance. The record is somewhat better with other accords. As of fall 1998, 83 percent of the members of the U.N. climate change convention had filed the national communications required under the treaty. And 83 percent of the parties to the Montreal Protocol on ozone depletion had reported data for 1996, although only 26 percent had submitted data for 1997.[5]

The mini-institutions set up by each treaty play a key role in the implementation process. At a minimum, each treaty spawns a conference of the parties (COP) and a secretariat. The COPs are regular meetings of treaty members; they provide an opportunity to strengthen the agreement and review problems in implementation. Secretariats are small offices set up to service these meetings of governments. Environmental conventions also commonly include scientific bodies, which provide advice on new scientific and technological information relevant to the implementation of the accord.[6]

Governments all too often give secretariats limited resources and authority. For instance, the secretariats generally do not have the wherewithal or authority to verify the information that governments are supposed to supply on implementation efforts. A typical secretariat has fewer than 20 staff and an annual budget of $2–11 million—a drop in the bucket compared with the budgets of U.S. federal agencies charged with implementing domestic environmental laws.[7]

A notable exception is the CITES (endangered species) Secretariat, which has been granted considerable powers by governments and has used them to positive effect. It can, for example, request information from countries about alleged lapses, and demand explanations from any it believes are falling short of meeting treaty obligations. Nevertheless, limited resources have prevented the secretariat from making full use of its authority.[8]

The scattered locations of the secretariats poses an added challenge. For instance, the secretariat for the ozone treaty is in Nairobi; the climate change treaty's office is in Bonn; and the biodiversity treaty secretariat is in Montreal. The U.N. Environment Programme (UNEP) has been granted control over some of these administrative groups; others report to different U.N. agencies. Centralizing these bodies under one roof would offer opportunities for the exchange of information and ideas, thereby making international environmental governance more efficient. In Agenda 21, governments cautiously endorsed the idea of centrally located convention secretariats, but in practice they have been reluctant to follow through on this. Although a small amount of consolidation has taken place in Geneva, further efforts in this direction have run afoul of the desires of individual countries to house these offices in order to enhance their own prestige.[9]

Although transparency is a powerful enforcement tool, in some cases stronger medicine is required. Trade restrictions can play an important role in encouraging countries to participate in international environmental accords, or to abide by those they have signed on to. But the use of trade levers as an enforcement tool is controversial, given the possible conflict with WTO rules. (See also Chapter 7.)

The use of unilateral environmental trade measures is particularly controversial. Such restrictions have been used in several cases to strengthen multilateral accords. Under a

U.S. law known as the Pelly Amendment, the government is authorized to impose trade sanctions against countries whose nationals are known to be diminishing the effectiveness of international fishery or wildlife agreements. Though the sanctions have rarely been invoked, the threat that they might be has helped to strengthen a number of international accords. For example, it helped secure the participation of Japan and Taiwan in the 1993 U.N. moratorium on destructive driftnet fishing, and also helped convince Japan to stop importing endangered sea turtles for use in jewelry and eyeglass frames, which is forbidden under CITES.[10]

Multilateral agreements themselves at times either stipulate or authorize trade restrictions. The Montreal Protocol on ozone depletion, for example, restricts signatories from trading in CFCs and products containing them with countries that have not joined in the accord. These provisions are widely credited with helping to bring about near universal participation in this landmark treaty. The Basel convention on hazardous waste export also restricts trade in wastes with countries that are not members of the accord. And CITES is empowered to recommend that members suspend wildlife trade with countries identified as out of compliance with the treaty's terms. It did just that regarding China, Italy, Taiwan, and Thailand in the early 1990s, and Greece in 1998. Where the sanctions have been imposed, as they were by the United States against Thailand in 1991 and against Taiwan in 1994, they have generally prompted stronger government enforcement of CITES.[11]

The conflicts between international trade rules and the provisions of some multilateral environmental agreements (MEAs) have been the subject of extensive discussion in recent years at the WTO's Committee on Trade and Environment. Although no country has ever lodged a formal WTO protest against an MEA, the potential for such challenges clearly exists. Even without one, worries about inconsistencies with WTO rules undoubtedly enter into the calculus of environmental treaty negotiators, dissuading them from using tools that would give teeth to environmental accords.[12]

One solution to this problem would be to amend the environmental exceptions to the WTO to clarify that trade measures taken pursuant to MEAs should be protected from challenge at the trade body. A precedent for this approach is provided by the North American Free Trade Agreement, which stipulates that in cases where its provisions conflict with those of three "grandfathered" environmental treaties (the Basel Convention, CITES, and the Montreal Protocol), the environmental treaty shall prevail. Although the European Commission once put forth a proposal roughly along these lines aimed at protecting MEAs from WTO challenge, it was not widely supported by other governments. The United States has in fact pushed for just the opposite approach in recent environmental treaty negotiations, where it has argued for clauses that clarify that nothing in the agreement should be construed as superseding pre-existing international obligations, including WTO rules.[13]

FINANCING CHALLENGES

Although the punitive approach embodied by penalties and sanctions has its place, it is not always appropriate or effective. Shortages of financial and technological resources, more than a lack of will, render many developing countries unable to comply with some treaty requirements. Thus a critical issue for the success of most treaties is whether adequate funding is made available to help developing countries make the investments required to meet the treaty's terms. The ozone treaty was the first environmental accord to create a sizable fund for this purpose. (See Chapter 6.)

With this model in mind, countries created the Global

Environment Facility (GEF) on an interim basis in 1991 to finance investments in preserving the global commons—the atmosphere, biological diversity, and international waterways. The GEF is a funding mechanism for both the biological diversity and the climate change treaties. In March 1994, governments agreed to make the GEF permanent, and to replenish it with $2 billion in new resources to be spent over four years; in 1998, countries committed another $2.8 billion to continue funding into the next century. As of June 1999, the facility had allocated some $2.5 billion in grants— 39 percent for biological diversity preservation, 36 percent for climate-related initiatives, 15 percent to protect international waters, 6 percent for projects aimed at heading off ozone depletion, and 4 percent for overarching projects.[14]

The GEF is an innovation in global governance, as it bridges the United Nations and Bretton Woods systems. Not wanting to create an entirely new institution, governments decided to make the GEF a joint undertaking of the U.N. Development Programme (UNDP), UNEP, and the World Bank. UNDP is charged primarily with carrying out technical assistance and capacity-building projects and with implementing a program of small grants to NGOs. UNEP is responsible for advancing environmental management at the regional and global levels, and with providing scientific and technical advice. And the World Bank is responsible for developing and implementing most investment projects, for mobilizing resources from the private sector, and for serving as the trustee for the GEF trust fund. In recent years, GEF has begun to work more with other partners in executing its projects, including regional development banks, NGOs, and the private sector.[15]

GEF's governing council employs an unusual "double majority" voting structure. Under this system, decisions are normally made by consensus. But in cases where this proves impossible and a matter is put to a vote, two consecutive tallies are required—the first on the basis of a one-nation, one-vote system similar to that used at the United Nations, and the second by a one-dollar, one-vote system comparable to that of the Bretton Woods institutions. This voting arrangement is intended to make the facility a joint undertaking of donors and recipients—a novel concept that offers a useful model for the governance of other international institutions.[16]

In April 1998, representatives of 119 member governments, 16 international organizations, and 185 NGOs gathered in New Delhi for the first full assembly of GEF members. On the table was an study of the fund's overall effectiveness that had been prepared by an independent team of consultants. The report raised several concerns about the GEF's performance, and offered recommendations for change.[17]

One controversial issue highlighted by the evaluation report is the facility's mandate to finance only the "incremental," or extra, costs to countries of investing in projects of global benefit, above and beyond the costs they would otherwise expect to encounter in their development. This requirement frustrates recipient countries, who resent the tendency of donor states to be more concerned about responding to abstract global threats than about addressing urgent local environmental needs. The incremental cost concept can also foster the false notion that activities such as preserving biological diversity, investing in renewable energy, and preventing coastal pollution are not worth pursuing out of national and local self-interest alone. To avoid the danger of serving as a disincentive for investing in projects that offer global and local benefits at the same time, the calculation of incremental costs is best viewed as a loose guiding concept rather than a precise operational procedure in the implementation of GEF projects.[18]

Given its limited resources, a key question for the GEF is how it can leverage these funds in support of broader efforts to reorient both national policies and other international financial flows in support of environmentally sound development strategies. One longstanding concern about the GEF is that merely by virtue of its existence the organization reduces pressure on the U.N. agencies and the Bank to integrate environmental issues broadly throughout their far larger overall lending programs. The evaluation report found reason for concern that this was in fact happening, particularly at the World Bank, which recently spent more in one year on carbon-emitting fossil fuel projects ($2.3 billion) than the GEF's entire replenishment for 1994–98.[19]

In order to maximize its influence, the GEF has been working over the last several years to increase its involvement with the private sector. In particular, it cooperated with the International Finance Corporation (IFC) to spearhead the creation of two private capital funds: a biodiversity fund for Latin America dubbed the Terra Capital Fund that has $15 million to finance sustainable forestry and agriculture programs and ecotourism projects, and a $100–240 million private capital fund promoting energy efficiency and renewable energy projects worldwide. Another promising joint GEF-IFC program channels funds through environmental NGOs, nonprofit venture capital firms, and other intermediaries to a range of small-scale, environmentally sound enterprises. Projects in renewable energy, energy efficiency, sustainable forestry and agriculture, and ecotourism are the funding targets. The program was originally capitalized at $4.3 million. It has now been expanded with an additional $16.5 million and will involve some 100 different projects when fully up and running.[20]

The Clean Development Mechanism (CDM) set up under the Kyoto Protocol is another creative effort to harness private capital for the implementation of international environmental commitments. Under the CDM, countries will be able to receive emissions credits for climate-friendly investments undertaken in developing countries that are in addition to those they would have undertaken in any case. There are many practical difficulties in certifying and monitoring these programs—problems similar to those encountered in putting into practice the "incremental cost" requirement for GEF projects. Governments are in the process of working out important details of the Clean Development Mechanism. Although the CDM holds promise as a way to harness private capital toward climate-protecting investments in the developing world, care needs to be taken in writing its rules to prevent it from being misused by governments and companies trying to evade reduction commitments.[21]

INSTITUTIONALIZING GLOBAL ENVIRONMENTAL PROTECTION

Although the proliferation of environmental treaties over the last few decades is an encouraging development, part of the price of success has been a measure of duplication and inefficiency. Rather than a cohesive system for the environment, what the world has is a patchwork quilt. This disorganized system needs to be streamlined if it is to become capable of reversing ecological decline.

The U.N. Environment Programme was intended to be the linchpin of international environmental cooperation when it was created in 1972 at the U.N. Conference on the Human Environment in Stockholm. At that time, the important role that independent environmental agencies play at the national level had yet to be demonstrated. Rather than creating a full-fledged U.N. environment agency, governments instead charged UNEP with catalyzing environmental activities throughout the U.N. system, including at much larger U.N. specialized agencies such as the Food and Agriculture Organization and the World Health Organization. Because UNEP was not to engage in projects of its own, governments decided that only a small staff was needed. UNEP's architects created an "environment fund" as a main tool through which the agency was to catalyze environmental initiatives throughout the United Nations.[22]

But governments failed to deliver on their promises of significant resources for the environment fund. In its first two decades, UNEP's total resources amounted to some $1 billion, less than UNDP's annual budget. UNEP's budget today—just under $100 million a year—is comparable to the budgets of some private environmental groups. UNEP is significantly smaller than most other U.N. agencies, and its resources pale in comparison to the lending programs of the World Bank and the International Monetary Fund. (See Table 9–2.) Another problem is UNEP's location in Nairobi, far from the other agencies it is theoretically coordinating.[23]

Despite these handicaps, UNEP has had its share of successes since 1972. It has played a central role in the negotiation of numerous environmental action plans and treaties, including a successful program that has developed action plans for the shared management of 14 regional seas, and the Montreal Protocol on ozone depletion. The organization's Division of Technology, Industry, and Economics, located in Paris, has been an important source of technical advice on waste-minimizing technologies and on the prevention of industrial accidents. In recent years, UNEP's regional offices have been strengthened, giving the agency a presence in Bangkok, Geneva, Mexico City, Manama (in Bahrain), and soon in Washington.[24]

UNEP's governing council clarified the program's role and mandate in a February 1997 Nairobi Declaration. Among the functions identified were monitoring and assessing global and regional environmental trends; further developing international environmental law, in part by encouraging coherence among the environmental conventions; promoting implementation of existing international agreements; better coordinating the environmental activities of the U.N. system as a whole; spearheading cooperation among various sectors of society involved with international environmental issues, including linking the scientific community with policymakers; and helping governments build environmental institutions. Since taking office in early 1998, UNEP Executive

TABLE 9–2

Estimated Expenditures and Staffing of Various International Agencies, 1998[1]

Agency	Staff (number)	Expenditure (million dollars)
World Bank	9,262	28,594
International Monetary Fund	2,196	27,495[2]
U.N. Development Programme	5,300	2,131
World Health Organization	3,498	900
U.N. Children's Fund	5,594	882
Food and Agriculture Organization	3,599	595
U.N. Education, Scientific, and Cultural Organization	2,200	405
International Atomic Energy Agency	2,221	274
International Labour Organisation	1,788	248
U.N. Population Fund	972	217
U.N. Industrial Development Organization	755	158[3]
U.N. Environment Programme	486	93
World Trade Organization	500	83[4]
World Meteorological Organization	246	56[4]
International Maritime Organization	300	28

[1]Expenditures reflect disbursement, lending, or budget figures, and are for 1998, except International Atomic Energy Agency (1997) and WTO (1999). Multiyear figures were divided evenly to obtain one-year equivalent. See endnote for more details. Includes professional and general service staff.   [2]Converted from SDR at the exchange rate of SDR1 = US$1.38, on 7 December 1999. [3]Does not include funding for technical cooperation.   [4]Converted from Swiss francs at the exchange rate SFr1 = US$0.6545, on 24 June 1999. SOURCE: See endnote 23.

Director Klaus Töpfer has taken steps to integrate these priorities into the organization's program.[25]

UNEP's recent reforms hold promise, but many observers argue for more far-reaching steps to raise the prominence of environmental issues within the United Nations. One idea now gaining political currency is to upgrade UNEP into a World Environment Organization (WEO) on a par with the WTO. Some will argue—as they did when UNEP was created in 1972—that establishing a strong environment agency would take pressure off the other U.N. agencies to integrate environmental considerations into their programs. But such integration is needed at the national level as well, and experience suggests that the existence of strong domestic environment agencies has promoted rather than discouraged it.[26]

Upgrading the status of environmental issues within the U.N. system is long overdue. Still, it is important that debates over form not distract from the ultimately far more important questions of function. In addition to UNEP's current roster of functions, a World Environment Organization could usefully serve as an umbrella for the current scattered collection of treaty bodies, just as domestic environment agencies oversee the implementation of national environmental laws. UNEP is already working to promote synergies and coordination between the environmental conventions. Bringing the treaty bodies under one roof could improve the opportunities for bargaining and facilitate NGO access. But in order for the WEO to have the clout it needs, the treaties themselves must stipulate strong enforcement capacities, and the institution would need to be endowed with sufficient financial resources to catalyze innovative programs.[27]

Although far from perfect, a precedent is provided by the International Labour Organisation (ILO), which constantly modifies and strengthens the hundreds of standards it has issued on concerns such as workplace safety and child labor. The ILO also reviews whether members are complying with

its standards and provides countries with technical assistance to help them with this task. It often generates enough
pressure in a first, investigative stage to bring an errant
country into line, making its second stage—a public hearing
to explain delinquency—unnecessary. Representatives from
both management and labor actually form part of the governing body of the ILO, through a unique tripartite system
in which they share equal standing with governments.[28]

THE UNITED NATIONS AND SUSTAINABLE DEVELOPMENT

Beyond a stronger U.N. environment agency, it is also essential that environmental concerns be integrated widely
throughout all U.N. activities. Many agencies are already
important players in the environmentally sustainable development arena.

At the Rio Earth Summit in 1992, governments entrust-
ed the U.N. Development Programme with promoting
"capacity building" for sustainable development by helping
countries design appropriate policies and strengthen the
domestic institutions required to implement them. Two
years later, UNDP created a Sustainable Energy and Environment Division (SEED) to consolidate the agency's environmental initiatives, including work on capacity building,
energy and atmosphere issues, the GEF, natural resources
management, and desertification. SEED is also charged with
infusing a concern for environmentally sound development
throughout the agency's programs.[29]

Numerous other U.N. agencies are also active on the
environment and sustainable development. The World
Meteorological Organization has made important contributions to improved understanding of the complexities of
climate science through its cosponsorship of the Intergovernmental Panel on Climate Change. The World Health

Organization promulgates air and water pollution guidelines that are considered the international norm. The U.N. Food and Agriculture Organization is actively involved in promoting sustainable agriculture projects and in protecting dwindling fisheries. And the U.N. Population Fund oversees implementation of the groundbreaking action plan produced in 1994 at the International Conference on Population and Development in Cairo.[30]

Another important U.N. player is the Commission on Sustainable Development (CSD). Created at the Earth Summit in 1992, the CSD serves as a forum where governments and nongovernmental participants review progress in implementing Agenda 21, share information about what works and what does not, and discuss impediments such as inadequate financial resources or lack of access to innovative technologies. The CSD was given the task of monitoring the activities of national governments, international organizations, and private actors.[31]

National and local governments alike have used the CSD to share information about successes and failures in implementing the Rio accords at national and local levels. Agenda 21 called on all nations to devise national sustainable development strategies, and by December 1999 a total of 140 had created national organizations or governmental units charged with implementing Agenda 21. There is also a growing movement worldwide to create sustainable cities and communities. The Toronto-based International Council for Local Environmental Initiatives is spearheading a campaign to promote the adoption of local Agenda 21s or similar undertakings. More than 2,000 cities in 64 countries— including Buga, Colombia; Quito, Ecuador; and Lahti, Finland—already have such initiatives under way.[32]

Governments have used the CSD to exchange views on contentious topics that cut across traditional issue dividing lines. For instance, the commission has considered the role of trade and finance in sustainable development, as well as the question of changing unsustainable production and consumption patterns. It is also working to encourage governments to develop and use sustainable development indicators to supplement traditional reliance on national income accounts. The CSD has proved useful as a launching pad for several important initiatives, including an intergovernmental panel on forests and a review of voluntary industry environmental initiatives.[33]

Perhaps most important, the CSD brings together a range of stakeholders on an annual basis to take stock of progress in putting the Rio agreements into practice. Since the first CSD session in 1993, the number of nongovernmental participants from around the world attending the annual forums has quadrupled, reaching some 800 in 1999. High-level government ministers, local officials, business organizations, farmers, and indigenous peoples, among others, have all participated. Together, these diverse players have helped push forward action on such issues as forests, oceans, production and consumption, tourism, trade, and finance.[34]

All these developments are to the good. Yet as the number of international environmental meetings proliferates, the ecological health of the planet continues to deteriorate. Almost 10 years after the Rio Earth Summit, a framework of international environmental governance has begun to emerge. The time has now come to move beyond the framework and construct the edifice itself.

CHAPTER 10

PARTNERSHIPS FOR THE PLANET

One of the great ironies of the demonstrations at the World Trade Organization (WTO) meeting in Seattle in late 1999 is that the massive protests against one form of globalization were facilitated by another dimension of the same process—the revolution in information and communications technologies. Just as these technologies have helped corporations go global, so have they facilitated powerful new forms of international activism by citizens' groups: "The Internet has become the latest, greatest arrow in our quiver of social activism," Mike Dolan, an organizer of the Seattle protests, told the *Los Angeles Times*.[1]

As the breakdown of the Seattle talks demonstrated, civil society has become strong enough to stop global economic negotiations in their tracks. But the more important question may be whether it can now harness that strength to build a new kind of global governance from the ground up. A nascent system of international environmental governance is already beginning to emerge from diverse quarters, proving that governance is no longer just for governments.

Reversing ecological decline in the early decades of the new century will require innovative partnerships between many different actors, including nongovernmental organizations (NGOs), businesses, governments, and international organizations.[2]

NGOs RISING

The last few decades have seen a flowering of nongovernmental environmental activism at the national and grassroots levels around the world. There are now tens of thousands of environmentally related NGOs, most of which have been formed since the 1980s. Growth has been particularly rapid in many parts of the developing world and in Eastern Europe, where democratization has opened up political space for NGOs and where mounting environmental problems have generated pressure for change.[3]

The number of NGOs working across international borders soared during the last century, climbing from just 176 in 1909 to more than 23,000 in 1998. (See Figure 10–1.) Environmental groups have risen steadily as a share of the total, climbing from just 2 percent of all transnational social change NGOs in 1953 to 14 percent in 1993.[4]

Some of these organizations—such as Friends of the Earth, Greenpeace, and the World Wide Fund for Nature (WWF)—are themselves international, representing global bases of support rather than parochial national interests. They answer to constituencies sizable enough to rival the populations of some nations: WWF has nearly 5 million members, and Greenpeace has 2.5 million. Empowered by e-mail and the Internet, NGOs have gradually organized themselves into a range of powerful international networks, such as the Women's Environment and Development Organization, the World Forum of Fish Workers & Fish Harvesters, and the Climate Action Network. (See Table 10–1.) Taken

FIGURE 10–1

Number of International NGOs, 1956–98

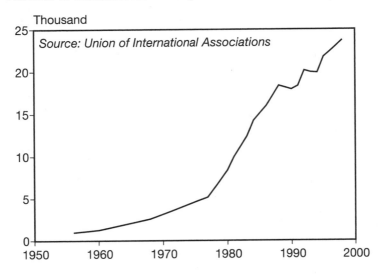

together, all this activity adds up to the creation of a bona fide global environmental movement that is rapidly becoming as influential at the international level as it already is within many countries.[5]

The last few years have seen some high-profile NGO successes. In one powerful demonstration of NGO clout, a coordinated campaign in 1998 to expose the environmental deficiencies of a proposed Multilateral Agreement on Investment (MAI) was successful in bringing the agreement to a halt, at least for the time being. Activists used the Internet to post and publicize the previously secret draft agreement, which aimed to lower obstacles to foreign investment just as the WTO eases the way for global trade. But NGOs exposed a number of provisions in the draft agreement that could have constrained governments' ability to minimize the environmental damage and social disruption of foreign investment projects. Some governments have been trying to revive

TABLE 10–1

Selected International Environmental NGO Networks

BirdLife International (created in 1922)

Global Reach: Links bird and habitat conservation organizations in 60 countries, with representatives in 27 additional countries.

Mission/Goal: To conserve bird species and their habitats and to identify priorities for bird and biodiversity conservation through research and data collection, grassroots efforts, and coordinated policies, campaigns, and programs.

International Federation of Organic Agriculture Movements (IFOAM) (1972)

Global Reach: Includes more than 500 member organizations in 100 countries.

Mission/Goal: To develop organic agriculture by working with producers, processors, and traders throughout the world.

Pesticide Action Network (PAN) (1982)

Global Reach: Links some 520 consumer, environment, health, labor, agriculture, and public interest groups worldwide.

Mission/Goal: To encourage the replacement of pesticides with ecologically sound alternatives and to promote more effective pest management through education, media, and advocacy.

International Rivers Network (IRN) (1985)

Global Reach: Links and mobilizes river activists worldwide via a global network of supporters, funders, advisors, interns, and volunteers.

Mission/Goal: To halt and reverse the degradation of river systems, to promote alternatives to dams and channeling, and to help protect and restore the well-being of the peoples, cultures, and ecosystems that depend on rivers, through research, analysis, and campaigning.

Climate Action Network (CAN) (1989)

Global Reach: Links more than 250 international groups and national organizations, organized into regional networks and represented by seven regional focal points.

Mission/Goal: To promote government and individual action to limit human-induced climate change to sustainable levels by coordinating information exchange, formulating policy options and position papers, and collaborating to promote effective NGO action on climate issues.

Women's Environment and Development Organization (WEDO) (1990)

Global Reach: Links a culturally and ideologically diverse group of women activists and experts worldwide.

Mission/Goal: To increase the participation of women in decisionmaking in public-policy institutions such as the United Nations and to achieve social, political, economic, and environmental justice through the empowerment of women.

Biodiversity Action Network (BIONET) (1993)

Global Reach: Currently composed of U.S. NGO members, but joins in cooperative projects worldwide and expects eventually to expand globally.

Mission/Goal: To advocate the effective implementation of the Biodiversity Convention worldwide, primarily through coordinated, joint NGO programs and information dissemination designed to catalyze governmental action.

International NGO Network on Desertification and Drought (RIOD) (1994)

Global Reach: Connects NGOs and community-based organizations world-wide via 16 focal points on six continents.

Mission/Goal: To improve the effectiveness of NGOs in their efforts to fight desertification through the exchange of information, expertise, and ideas.

World Forum of Fish Workers & Fish Harvesters (WFF) (1995)

Global Reach: Links fish harvesters, fish workers, and women from artisanal and small-scale fishing communities via regional councils on six continents.

Mission/Goal: To protect the rights and livelihood of fish workers while also promoting sustainable fishing methods and greater industry and government compliance with international fisheries agreements.

International POPs Elimination Network (IPEN) (1997)

Global Reach: Currently active on six continents; expects to gain the endorsement and participation of hundreds of NGOs worldwide.

Mission/Goal: To facilitate information exchange and to support the activities of members who share a common interest in achieving the global elimination of persistent organic pollutants.

SOURCE: See endnote 5.

the negotiations under the umbrella of the WTO, but the outcome of the Seattle meeting suggests that this is unlikely to happen anytime soon.[6]

Another genre of NGO activism is aimed directly at corporate decisionmakers rather than at international diplomats. In one example of this approach, Greenpeace seized headlines in 1995 when it successfully mobilized a consumer boycott of Shell gas stations, forcing the gargantuan multinational company to give up its plans for dumping the *Brent Spar*, an abandoned oil rig, in the North Atlantic Ocean. A few years later, MacMillan Bloedel, one of Canada's largest forest products companies, agreed to stop clearcutting in British Columbia after a campaign by Greenpeace and other groups to boycott products made from clearcut timber convinced the company that the environmental stigma of such a form of logging could devastate its European market.[7]

Monsanto and other large producers of transgenic seeds now find themselves in a similar position, as organized opposition to genetically modified organisms (GMOs) by European environmentalists has stymied their plans to introduce genetically modified foodstuffs in Europe. The waves of opposition have rippled back across the Atlantic, and U.S. grain exporters began in late 1999 to ask farmers to grow segregated non-GMO crops. In October 1999, Monsanto Chairman Robert Shapiro addressed his foes at a Greenpeace business conference in London, where he admitted that the company had not consulted sufficiently with its critics, and pledged to enter into such discussions "openly, honestly and non-defensively" in the future.[8]

Perhaps the strongest proof of the growing strength of citizens' movements is the seriousness with which the international business community now views them. A 1997 report by the Control Risks Group, a London-based firm that advises businesses on political and security risks, discusses the need for companies to obtain a social "license to operate," and describes "the pressure on companies, wherever they operate, to adopt the highest international environmental, labour and ethical standards." According to the report, "heightened international scrutiny means that perceived transgressors truly have 'no hiding place'."[9]

OPENING UP INTERNATIONAL INSTITUTIONS

International treaties and institutions have traditionally been viewed as compacts among sovereign nations. Individual citizens were granted no formal role in the international legal system, but were instead expected to make their voices heard by influencing the policies of their own national governments, thereby affecting the positions these governments advocate in international forums.[10]

But this theory is increasingly breaking down. Both environmental organizations and business groups now exert a direct and powerful influence in a broad range of international negotiations and institutions. U.N. conferences and the sessions in which environmental treaties are negotiated are all routinely attended by scores of NGOs from all over the world. International institutions such as the U.N. Commission on Sustainable Development (CSD), the Global Environment Facility, the World Bank, and the World Trade Organization are the subject of intense NGO interest and scrutiny. All these organizations have by now developed procedures—some more participatory than others—for interacting with the nongovernmental community.[11]

Despite the recent growth in interaction between NGOs and international institutions, the relationship is often strained. Citizens' groups working at the global level face formidable obstacles. There is as yet nothing resembling an elected parliament at the international level. Although the

United Nations has begun to experiment with occasional public hearings on topics of special concern, these continue to be rare events. No formal provisions are made for public review or comment on international treaties, nor is there a mechanism for bringing citizen suits at the World Court. International negotiations are often closed to public participation, and access to documents of critical interest to the public is generally restricted.[12]

Officials and government representatives, for their part, sometimes grow frustrated with NGOs. The confrontational tactics of some groups, such as the effort to delay the opening of the WTO meeting in Seattle, antagonize government officials. And the growing role of NGOs in international forums raises difficult issues of accountability. NGOs, unlike democratic national governments, cannot claim the legitimacy conferred by the ballot box. Their sources of credibility and support are more complex, deriving from factors such as unique knowledge or experience.[13]

Much of the NGO activity in international environmental policymaking dates to the June 1992 Earth Summit, which was a watershed for the democratization of global environmental governance. The 20,000 concerned citizens and activists who attended the Rio conference from around the world outnumbered official representatives by at least two to one.[14]

The Agenda 21 action plan produced at the Earth Summit encourages the democratization of international policymaking by focusing on the important role of "major groups" (including citizens' groups, labor unions, farmers, women, business interests, and others) and by endorsing the need to make information freely and widely available. In an important precedent, the Commission on Sustainable Development based its rules for NGO participation on the liberal regulations that were in effect for the Rio conference. As a result, hundreds of groups are authorized to observe CSD deliberations and make selective interventions. An international NGO Steering Committee promotes collaboration among these groups and facilitates interaction with the CSD secretariat as well as with governments.[15]

Daily newsletters produced by citizens' organizations, including *Eco* and the *Earth Negotiations Bulletin*, have now become mainstays of the international negotiating process. Widely read by official delegates and NGOs alike during international meetings, they reveal key failures in negotiations and prevent the obscure language of international diplomacy from shielding governments from accountability for their actions.[16]

The business community is also becoming a growing force in international environmental negotiations—for both good and ill. Its strong presence is now felt in many different international environmental forums, including those on climate change, biological diversity loss, and the control of persistent organic pollutants.

In the climate change negotiations, a number of U.S.-based companies, including coal and oil businesses, that maintain they would suffer under the treaty have participated in the Global Climate Coalition, a group that opposes the Kyoto Protocol. But other business groups are enthusiastically in favor of a strong accord. Business councils for sustainable energy have been formed over the last several years in Australia, Europe, and the United States; they include appliance manufacturers and renewable energy, energy services, cogeneration, and natural gas companies that have calculated that a strong climate treaty would help their bottom lines. The insurance industry has also become a convert to the cause, as it is worried that extreme weather disturbances caused by climate change could translate into steeply rising claims. (See Chapter 6.)[17]

Parliamentarians are another potentially powerful group of players on the global stage. The Global Legislators Organization for a Balanced Environment (GLOBE) was founded in 1989 to promote cooperation between parliamentarians on global environmental issues. It now has more than 700 members from more than 100 countries, with regional affiliates in Brussels, Capetown, Moscow, Tokyo, and Washington, DC. GLOBE has been active in a range of international policy debates, including those on climate change and the WTO.[18]

Despite the many advances that NGOs have made at the United Nations over the last several years, frustrations remain. One problem is the scarcity of opportunities to participate in the activities of standing U.N. bodies, including the General Assembly and the Security Council. And security restrictions imposed on NGO access to the U.N. premises over the last few years have made it increasingly difficult to lobby delegates directly. The New York–based Global Policy Forum charges that despite Secretary-General Kofi Annan's lofty speeches about the role of NGOs, his Secretariat has restricted their involvement through new rules, regulations and fees.[19]

Pressure is growing for the United Nations to take bold steps to formalize the growing importance of NGOs. Some proposals envision creating a new assembly within the world body where the views of the people of the world could be more directly represented than under the current system. One model for such an assembly is the directly elected European Parliament. A more feasible approach might be to create a body composed of representatives of national parliaments—perhaps as a transition to a full-fledged peoples' assembly. A "Millennium Forum" at the United Nations in May 2000 will bring together NGO representatives from around the world and other representatives of civil society to consider how the sector's energies can best be harnessed to confront the global challenges of the twenty-first century. The forum will consider, among other items, whether a new organizational structure is needed to aid citizen participation in global decisionmaking.[20]

As the outcry in Seattle made clear, bold actions are also needed at global economic institutions to make them more open and accountable. Information and documents at these organizations are often tightly guarded, and negotiations between governments are completely closed to observers, with no NGO newsletters offering blow-by-blow accounts of who said what to whom.

President Bill Clinton told delegates in Seattle that "if the WTO expects to have public support grow for our endeavors, the public must see, and hear, and...actually join in the deliberations. That's the only way they can know the process is fair and know their concerns were at least considered." Despite this plea, many countries resisted making the WTO's operations more transparent. Before the talks broke down in Seattle, some limited progress was made on allowing more documents to be declassified. But no consensus was reached on opening the controversial dispute resolution panels to public observation, or on allowing NGOs to submit "friend of the court" briefs.[21]

The World Bank, in contrast, has taken some important steps over the last several years to make its activities more accountable. In the early 1990s, the Bank adopted an information disclosure policy that, while by no means perfect, has made many more Bank documents publicly available. And the creation of an independent inspection panel in 1994 set an important precedent by providing an impartial forum where board members or private citizens can raise complaints about projects that violate the Bank's own policies, rules, and procedures. The International Finance Corporation has recently appointed an environmental and social Compliance Advisor/Ombudsman to serve a similar function for private-sector projects.[22]

Even the notoriously secretive International Monetary

Fund has moved to make more documents available to the public. But this organization still has a long way to go in learning how to listen to society at large when designing its lending programs.[23]

FORGING A NEW GLOBAL COMPACT

In January 1999, U.N. Secretary-General Kofi Annan addressed the World Economic Forum in Davos, Switzerland, an annual gathering of corporate titans and other members of the global elite. While acknowledging that globalization was now a fact of life, Annan noted that the spread of global markets was outpacing the ability of societies and their political systems to adjust to them, "let alone to guide their course." He called on the business leaders present in Davos to work with the United Nations to forge a new global compact that would "embrace, support and enact a set of core values in the areas of human rights, labour standards, and environmental practices."[24]

In the environmental realm, several business initiatives along these lines are already well under way. Many multinational corporations already adhere to roughly uniform environmental policies and standards throughout their worldwide operations. And a number of international industry groups have now crafted voluntary codes of environmental conduct; many of them call for companies to approximate the standards of their home countries wherever they do business, which is more practical in any case than operating in numerous different ways around the world. Meeting international environmental criteria also allows companies to trumpet green credentials, which are of growing value in the international marketplace. An additional impetus for maintaining strict internal corporate environmental policies is a desire to avoid adverse publicity—as well as recognition of the growing tendency for costly lawsuits to be filed in home-country courts for alleged environmental transgressions overseas.[25]

Over the last few years, some 10,000 companies worldwide, many from developing countries, have become certified under the voluntary environmental management guidelines forged by the Geneva-based International Organization for Standardization, a worldwide federation of national standards-setting bodies. The first set of standards in this "ISO 14000" series was finalized in the fall of 1996. It covers internal management and auditing procedures—how, for instance, a company should monitor its pollution. These management guidelines are not to be confused with actual performance standards that would specify, for example, what levels of pollution would be acceptable. But they are nonetheless a useful tool.[26]

Another type of international standard setting is embodied in the numerous independent eco-labeling initiatives now beginning to take hold. The organic agriculture community was a pioneer in this field. In the early 1970s, it came together through the International Federation of Organic Agriculture Movements to lay out the conditions that farmers must meet in order to claim organic credentials. More recently, the Forest Stewardship and Marine Stewardship Councils were formed to devise criteria for sustainable timber and fish harvesting. (See Chapters 2 and 4.) One strength of these efforts is the diverse range of stakeholders they bring to the table, including businesses, NGOs, and often governments and international organizations. In effect, these initiatives are defining sustainability criteria for the global economy.[27]

Although globalization is reducing the ability of governments to regulate activities within their borders, it is also opening the way for a number of innovative public-private partnerships such as those just described. Political economist Wolfgang Reinicke argues that such collaborations are planting the seeds of a new system of international governance based on global public policy networks among diverse actors, including international organizations, business, labor, and NGOs. Although such partnerships will not substitute for governments in the decades ahead, they could play a key role in helping to bring together the various players needed to solve diverse aspects of the planet's ecological predicament.[28]

Thirty years ago, photographs of Earth taken from space by the Apollo expeditions indelibly impressed on all who saw them that the planet, while divided by political boundaries, is united by ecological systems. These photos helped inspire the first Earth Day, which in turn motivated numerous countries to pass environmental laws and create environmental ministries. This year, the world will celebrate Earth Day 2000. A comparable groundswell is needed in support of the innovations in global governance that will be required to safeguard the health of the planet in the new millennium.

NOTES

CHAPTER 1. ONE WORLD?

1. Suzanne Pardee, "Demonstrators Swarm WTO Ministerial Meeting," *Environment News Service*, <www.ens.lycos.com/ens/nov99/1999L-11-30-01.html>, 30 November 1999; Bill McKibben, "Making Sense of the Battle in Seattle," 1 December 1999, available at <www.gristmagazine.com/grist/maindish/mckibben 120199.stm>; Joseph Kahn and David E. Sanger, "Trade Obstacles Unmoved, Seattle Talks End in Failure," *New York Times*, 4 December 1999.

2. Steven Pearlstein, "Trade Theory Collides With Angry Reality," *Washington Post*, 3 December 1999.

3. For different takes on the meaning of globalization, see, for example, Wolfgang H. Reinicke, *Global Public Policy* (Washington, DC: Brookings Institution Press, 1998); David C. Korten, *When Corporations Rule the World* (West Hartford, CT: Kumarian Press, 1995); George Soros, *The Crisis of Global Capitalism* (New York: PublicAffairs, 1998); and Thomas L. Friedman, *The Lexus and the Olive Tree* (New York: Farrar, Straus and Giroux, 1999).

4. Table 1–1 based on the following sources: exports of goods from a Worldwatch data series derived from export values and the export unit value index (EUVI) supplied by Neil Austriaco,

Research Assistant, Statistics Division, International Monetary
Fund (IMF), e-mails to Lisa Mastny, Worldwatch Institute, 14
December 1999; services data from Andreas Maurer, World Trade
Organization (WTO), e-mail to Lisa Mastny, Worldwatch Institute,
20 January 1999, deflated using the EUVI; foreign direct invest-
ment (FDI) inflows from Lizanne Martinez, statistical assistant,
U.N. Conference on Trade and Development (UNCTAD), e-mail to
Lisa Mastny, 30 June 1999, and from UNCTAD, "Global Foreign
Direct Investment Boomed in 1998, Fuelled By Mergers and
Acquisitions," press release (Geneva: 22 June 1999); capital flows
to developing countries from World Bank, *Global Development
Finance 1999* (Washington, DC: 1999); number of transnational
corporations (TNCs) from Joshua Karliner, *The Corporate Planet*
(San Francisco, CA: Sierra Club Books, 1997), and from UNC-
TAD, *World Investment Report 1998* (New York: U.N., 1998); TNC
sales from "Worldbeater, Inc.," *The Economist*, 18 January 1998,
and from UNCTAD, *World Investment Report 1998*, op. cit. this
note; world shipping from UNCTAD, *Review of Maritime Transport*
(Geneva: various years), and from UNCTAD, "World Seaborne
Trade Slows Down," press release (Geneva: 22 February 1999);
unit cost of shipping from U.N. Development Programme
(UNDP), *Human Development Report 1999* (New York: 1999); air
transport data from Attilio Costaguta, Chief, Statistics and Economic Analysis Section, International Civil Aviation Organization
(ICAO), e-mail to Lisa Mastny, Worldwatch Institute, 2 November
1998, and from ICAO, "World Airline Passenger Traffic Growth To
Pick Up Through To 2001" (Montreal: 21 July 1999); air revenue
from UNDP, op. cit. this note; tourism growth from Rosa Songel,
Department of Statistics and Economic Measurement of Tourism,
World Tourism Organization, e-mail to Lisa Mastny, Worldwatch
Institute, 25 November 1999; 2 million crossing borders daily
from Institute of Medicine, *America's Vital Interest in Global Health*
(Washington, DC: National Academy Press, 1997); international
refugees from U.N. High Commissioner for Refugees, "UNHCR by
Numbers," <www.unhcr.ch/un&ref/numbers/table1.htm>, viewed
6 January 1999; total refugees from Michael Renner, "Refugee
Numbers Drop Again," in Lester R. Brown, Michael Renner, and
Brian Halweil, *Vital Signs 1999* (New York: W.W. Norton & Com-

pany, 1999); phone lines from International Telecommunication Union (ITU), *World Telecommunication Indicators on Diskette* (Geneva: 1996), and from ITU, "Telecommunications Industry at a Glance," <www.itu.int/ti/industryoverview/index.htm>, viewed 14 December 1999; phones in developing countries from ITU, *Indicators on Diskette*, op. cit. this note, from ITU, *Challenges to the Network: Telecoms and the Internet* (Geneva: September 1997), and from ITU, *World Telecommunication Development Report* (Geneva: 1999); phone cost from UNDP, op. cit. this note; Internet growth (measured by number of host computers) from Network Wizards, "Internet Domain Surveys, 1981–1999," <www.nw.com>, updated January 1999; 1998 users is a Worldwatch estimate of individuals who use the Internet on a weekly basis, based on Nua Ltd., "How Many Online?" <www.nua.ie>, updated February 1999, and on Computer Industry Almanac, Inc., *Internet User Forecast 1990–2005* (Arlington Heights, IL: 1999); on-line spending from Barry Parr, Internet and e-commerce analyst, International Data Corporation, Framingham, MA, discussion with Payal Sampat, Worldwatch Institute, 23 February 1999; Internet access from Computer Industry Almanac, Inc., op. cit. this note; unit cost of computing from UNDP, op. cit. this note; number of nongovernmental organizations from Union of International Associations, *Yearbook of International Organizations 1998–99* (Munich: K.G. Saur Verlag, 1999); survey of 22 nations from Lester M. Salamon and Helmut K. Anheier, *The Emerging Sector Revisited: A Summary of Initial Estimates* (Baltimore, MD: The Johns Hopkins University Center for Civil Society Studies, 1998).

5. Jared Diamond, *Guns, Germs, and Steel* (New York: W.W. Norton & Company, 1997); Clive Ponting, *A Green History of the World* (New York: Penguin Books, 1991); on earlier periods of globalization compared with today's, see Friedman, op. cit. note 3.

6. Figures on growth in world economy derived from Worldwatch data series for gross world product based on purchasing power parity, which is derived from Angus Maddison, *Monitoring the World Economy, 1820–1992* (Paris: Organisation for Economic Co-operation and Development (OECD), 1995), from Angus Maddison, *Chinese Economic Performance in the Long Run* (Paris: OECD, 1998), and from IMF, *World Economic Outlook and Interna-*

tional Capital Markets Interim Assessment, October 1999 (Washington, DC: 1999); exports and Figure 1–1 based on a Worldwatch data series derived from Austriaco, op. cit. note 4; share of exports in world economy based on these two series, versions of which are published in Brown, Renner, and Halweil, op. cit. note 4.

7. FDI from Martinez, op. cit. note 4, and from UNCTAD, "Global Foreign Investment Boomed in 1998," op. cit. note 4, with figures deflated using U.S. GNP Implicit Price Deflator, from U.S. Department of Commerce, *Survey of Current Business*, July 1999; TNCs from Karliner, op. cit. note 4, and from UNCTAD, op. cit. note 4; households from Investment Company Institute, "Mutual Fund Developments in 1998," in *Mutual Fund Factbook* (Washington, DC: May 1999); $16 billion from John Rea, "U.S. Emerging Market Funds: Hot Money or Stable Source of Investment Capital?" Investment Company Institute, December 1996; 1996 from Investment Company Institute, *Mutual Fund Factbook* (Washington, DC: 1997).

8. Ted C. Fishman, "The Joys of Global Investment," *Harper's Magazine*, February 1997.

9. Arctic contamination from Arctic Monitoring and Assessment Programme, *Arctic Pollution Issues: A State of the Arctic Environment Report* (Tromso, Norway: 1997).

10. Wood consumption figures from U.N. Food and Agriculture Organization (FAO), *FAO Forest Products Yearbook 1983–1994* (Rome: 1996); sixfold increase in paper use calculated from 46 million tons in 1950 from International Institute for Environment and Development, *Towards a Sustainable Paper Cycle* (London: 1996), and from 299 million tons in 1997 from Miller Freeman, Inc., *International Fact and Price Book 1999* (San Francisco: 1998); fivefold increase in fish consumption from FAO, *Yearbook of Fishery Statistics: Catches and Landings* (Rome: various years); water consumption from Sandra Postel, *Last Oasis*, rev. ed. (New York: W.W. Norton & Company, 1997); grain consumption figures from FAO, *The State of Food and Agriculture 1995* (Rome: 1995); steel use figures from International Iron and Steel Institute, *Steel Statistical Yearbook* (Brussels: various years); fossil fuel burning figures based on T.A. Boden, G. Marland, and R.J. Andres, *Estimates of Global, Regional and National Annual CO_2 Emissions From Fossil*

Fuel Burning, Hydraulic Cement Production, and Gas Flaring: 1950–92, electronic database, Carbon Dioxide Information Analysis Center, Oak Ridge National Laboratory, Oak Ridge, TN, December 1995, and on British Petroleum, *BP Statistical Review of World Energy 1997* (London: Group Media & Publications, 1997); population figures for 1950 from U.S. Bureau of the Census, International Programs Center (IPC), *International Data Base*, electronic database, <www.census.gov/ipc/www/idbnew.html>, updated 28 December 1998, and for 1999 from U.S. Bureau of the Census, IPC, "World POPClock Projection," <www.census.gov/cgi-bin/ipc/popclockw>, viewed 23 November 1999.

11. Figure 1–2 and carbon dioxide concentrations from Seth Dunn, "Carbon Emissions Dip," in Brown, Renner, and Halweil, op. cit. note 4; warning on mass extinction from American Museum of Natural History, "National Survey Reveals Biodiversity Crisis—Scientific Experts Believe We Are in Midst of Fastest Mass Extinction in Earth's History," press release (New York: 20 April 1998); plant extinction data from Kerry S. Walter and Harriet J. Gillett, eds., *1997 IUCN Red List of Threatened Plants* (Gland, Switzerland: World Conservation Union (IUCN), 1997); mammal figures from Jonathan Baillie and Brian Groombridge, eds., *1996 IUCN Red List of Threatened Animals* (Gland, Switzerland: IUCN, 1997).

12. Mathis Wackernagel et al., "National Natural Capital Accounting with the Ecological Footprint Concept," *Ecological Economics*, vol. 29 (1999); Mathis Wackernagel and Alejandro Callejas, "The Ecological Footprint of 52 Nations (1995 data)," Redefining Progress, available at <www.rprogress.org>.

13. Dani Rodrik, "The Global Fix," *New Republic*, 2 November 1998.

CHAPTER 2. NATURE UNDER SIEGE

1. Warning on mass extinction from American Museum of Natural History, "National Survey Reveals Biodiversity Crisis—Scientific Experts Believe We Are in Midst of Fastest Mass Extinction in Earth's History," press release (New York: 20 April 1998).

2. Figure 2–1 based on the following sources: map of biodiversity hotspots and tropical wilderness areas provided by Stephen

Nash, State University of New York–Stony Brook, for Conservation International, e-mail to Liz₎Doherty, Worldwatch Institute, 22 October 1999; logging concessions in Cameroon, Cambodia, Madagascar, Guyana, Republic of Congo, Suriname, Côte d'Ivoire, Papua New Guinea, and Amazon basin from Cheri A. Sugal and Russell A. Mittermeier, "Transnational Logging Investments in the Major Tropical Wilderness Areas," *CI Policy Brief* (Washington, DC: Conservation International, summer 1999); U.S. oil development from Yereth Rosen, "Alaska's Nature Advocates Fret Over Oil Field 'Sprawl' in Arctic," *Christian Science Monitor*, 9 April 1998; Canadian mining from Julian Pettifer, "Arctic Diamonds," *Review* (Rio Tinto), June 1999; Baja saltworks from "Locals Tackle Ecologists Over Mexico Saltworks," *Reuters*, 6 July 1999; Mexican plantations from John Ross, "Big Pulp vs. Zapatistas," *Multinational Monitor*, April 1998; Mexican mining from Koren Capozza, "Mexican Town Protests Canadian Mining Operation," *Environment News Service*, <ens.lycos.com/ens/may99/1999L-05-14-03.html>, 14 May 1999, and from Border Ecology Project, "Environmental/Social Impacts of Multinational Mining Investment in Sonora, Mexico" (Bisbee, AZ: 26 October 1995); Colombian oil from Steven Dudley and Mario Murillo, "Oil in a Time of War," *NACLA Report on the Americas*, March-April 1998; Ecuadorean oil from "Texaco vs Ecuadorean Tribes," *Environment News Service*, <ens.lycos.com/ens/feb99/1999L-02-03-03.htm>, 3 February 1999; Peru from "The Mining Elite," *Tomorrow*, July/August 1998; Bolivia pipeline from Nancy Dunne, "OPIC Set to Approve Bolivian Pipeline Loan," *Financial Times*, 15 June 1999; Guyana mining from Forest Peoples Programme, "Guyana Government Grants 5.1 Million Acre Mining Concession on Indigenous Lands," *Guyana Information Update* (Moreton-in-Marsh, U.K.: 9 November 1998); Suriname mining from Forest Peoples Programme, "Maroon Community Petitions Suriname Government About the Operations of a US-Owned Bauxite Company," *Suriname Information Update* (Moreton-in-Marsh, U.K.: 17 September 1998); Chile mining from James Langman, "Environment Restriction on Mining in Chile Weak, Say Experts, United Nations Report," *International Environment Reporter*, 27 May 1998; Chile forest products from James Langman, "Eco-Activists Fight Projects Doubling

Chile Forest Exports," *Christian Science Monitor*, 4 March 1999;
Mediterranean from World Wide Fund for Nature, "WWF Calls
for Better Protection of Valuable Mediterranean Forests," press
release (Gland, Switzerland: 1 July 1999); Ghana mining from
Charles Abugre and Thomas Akabzaa, "Mining Boom—A Gain for
Africa?" Third World Network, <www.twnside.org.sg/souths/twn/
title/boom-cn.htm>, viewed 9 December 1998; Cameroon-Chad
pipeline from "World Bank Completes Chad Oil Study, Waits for
Consortium," *Reuters*, 12 November 1999; Philippines from Deb-
orah Johansen, "Digging Upwards," *Tomorrow*, July/August 1998;
Papua New Guinea mining from "Mine Closure Threatens Papua
Economy Rescue Plans," *Reuters*, 13 August 1999; Papua New
Guinea oil from Jared Diamond, "Paradise and Oil," *Discover*,
March 1999; Indonesia mining from "Freeport Amends Motion in
$6 Billion Suit Alleging Firm's Mine Degraded Environment,"
International Environment Reporter, 12 June 1996; Indonesia plan-
tations from Japan Paper Association, *Pulp & Paper Statistics 1999*
(Tokyo: 1999); New Zealand logging from Stephanie Nall and
William Armbruster, "US Forest Products Companies Eye New
Zealand," *Journal of Commerce*, 19 May 1997; Russia oil and gas
from Bruce Forbes, "The End of the Earth," *Wild Earth*, fall 1999.

3. Stuart L. Pimm, "The Value of Everything," *Nature*, 15 May
1997; gross world product of $42 trillion is a Worldwatch estimate
for 1998 derived from International Monetary Fund (IMF) figures,
based on Lester R. Brown, "Global Economic Growth Slows," in
Lester R. Brown, Michael Renner, and Brian Halweil, *Vital Signs
1999* (New York: W.W. Norton & Company, 1999).

4. Species loss from Joby Warrick, "Mass Extinction Under-
way, Majority of Biologists Say," *Washington Post*, 21 April 1998,
and from American Museum of Natural History, op. cit. note 1;
ecosystem transformation from Janet Abramovitz, "Ecosystem
Conversion Spreads," in Lester R. Brown, Michael Renner, and
Christopher Flavin, *Vital Signs 1997* (New York: W.W. Norton &
Company, 1997).

5. Convention on Biological Diversity (CBD) Clearing-House
Mechanism, "Text of the Convention on Biological Diversity
(1992)," <www.biodiv.org/chm/conv/default.htm>, viewed 24
November 1999; ratification figure from United Nations, *The Unit-*

ed Nations Treaty Collection, electronic database, <www.un.org/Depts/Treaty>, viewed 2 December 1999.

6. Forest services from Norman Myers, "The World's Forests and Their Ecosystem Services," in Gretchen C. Daily, ed., *Nature's Services: Societal Dependence on Natural Ecosystems* (Washington, DC: Island Press, 1997); nearly half lost from Dirk Bryant, Daniel Nielsen, and Laura Tangley, *The Last Frontier Forests* (Washington, DC: World Resources Institute (WRI), 1997); 14 million hectares based on estimates in U.N. Food and Agriculture Organization (FAO), *State of the World's Forests 1999* (Rome: 1999).

7. Janet N. Abramovitz, *Taking a Stand: Cultivating a New Relationship with the World's Forests*, Worldwatch Paper 140 (Washington, DC: Worldwatch Institute, April 1998); countries where exports exceed domestic consumption from Nigel Sizer, David Downes, and David Kaimowitz, "Tree Trade: Liberalization of International Commerce in Forest Products: Risks and Opportunities," *Forest Notes* (WRI and Center for International Environmental Law), November 1999; growth of plywood exports in Indonesia and Malaysia are volume figures from FAO, *FAOSTAT Statistics Database*, electronic database, <apps.fao.org>, viewed 22 October 1999.

8. Nigel Dudley, Jean-Paul Jeanrenaud, and Francis Sullivan, "The Timber Trade and Global Forest Loss," *Ambio*, May 1998; frontier forests from Bryant, Nielsen, and Tangley, op. cit. note 6.

9. Figure 2–2 from FAO, op. cit. note 7, viewed 22 October 1999. All data deflated using the U.S. GNP Implicit Price Deflator provided in U.S. Department of Commerce, *Survey of Current Business*, July 1999; factors behind future growth from Victor Menotti, "Forest Destruction and Globalization," *The Ecologist*, May/June 1999.

10. Developing-country exports from FAO, op. cit. note 7; share of global loss from three countries from FAO, op. cit. note 6.

11. Scope of illegal trade from Dudley, Jeanrenaud, and Sullivan, op. cit. note 8, and from Rob Glastra, ed., *Cut and Run: Illegal Logging and Timber Trade in the Tropics* (Ottawa: International Development Research Centre, 1999); Mexico from "Mexico's Monarch Butterflies in Danger," *Environmental News Network*, <www.enn.com/enn-subsciber-news-archive/1999/02/021199/

monarch.asp>, 11 February 1999.

12. United Nations Conference on Trade and Development (UNCTAD), *World Investment Report 1992* (New York: United Nations, 1992); Côte d'Ivoire figure from Nigel Dudley, Jean-Paul Jeanrenaud, and Francis Sullivan, *Bad Harvest: The Timber Trade and the Degradation of the World's Forests* (London: World Wide Fund for Nature and Earthscan, 1995); role of Japanese companies in 1970s and 1980s from Joshua Karliner, *The Corporate Planet: Ecology and Politics in the Age of Globalization* (San Francisco: Sierra Club Books, 1997).

13. Asian companies from Greenpeace International, *Logging the Planet: Asian Companies Marching Across Our Last Forest Frontiers* (Amsterdam: 1997), and from Carlos Sergio Figueiredo Tautz, "The Asian Invasion: Asian Multinationals Come to the Amazon," *Multinational Monitor*, September 1997; sale of logging rights from Sugal and Mittermeier, op. cit. note 2; record of timber companies from Nigel Sizer and Richard Rice, *Backs to the Wall in Suriname: Forest Policy in a Country in Crisis* (Washington, DC: WRI, April 1995).

14. Mexico from John Ross, "Treasure of the Costa Grande," *Sierra*, July/August 1996; Brazil from Ashley Mattoon, "Paper Forests," *World Watch*, March/April 1998; Chile from Langman, "Eco-Activists Fight Projects," op. cit. note 2; China from "APRIL Fine Paper's Changshu Mill Starts Production," *pponline.com Daily News*, <www.pponline.com>, viewed 25 March 1999; Indonesia from Lesley Potter and Justin Lee, *Tree Planting in Indonesia: Trends, Impacts and Directions*, Occasional Paper No. 18 (Bogor, Indonesia: Center for International Forestry Research, December 1998); for Argentina, and for a broader discussion of trends in the industry, see Janet N. Abramovitz and Ashley T. Mattoon, *Paper Cuts: Recovering the Paper Landscape*, Worldwatch Paper 149 (Washington, DC: Worldwatch Institute, December 1999); current Japanese investment from Japan Paper Association, op. cit. note 2.

15. Requirements of proposed agreement from "U.S. Says Trade Plan Won't Hurt Forests," *Reuters*, 4 November 1999; U.S. report from Office of the U.S. Trade Representative and White House Council on Environmental Quality, *Accelerated Tariff Liberalization in the Forest Products Sector: A Study of the Economic and*

Environmental Effects (Washington, DC: November 1999). For general discussions of the expected impact of trade liberalization on forests, see David Kaimowitz, "The Potential Environmental Impacts of Trade Liberalisation in Forest Products," *BRIDGES Weekly Trade News Digest*, July/August 1999, and Sizer, Downes, and Kaimowitz, op. cit. note 7.

16. Earthjustice Legal Defense Fund and Northwest Ecosystem Alliance, *Our Forests at Risk: The World Trade Organization's Threat to Forest Protection* (Seattle, WA: Earthjustice Legal Defense Fund, 1999); nontariff trade barriers from Forest Research, "Draft Study of Non-Tariff Measures in the Forest Products Sector in the APEC Economies, Part II: Inventory," study prepared for the Asia-Pacific Economic Cooperation Secretariat, August 1999.

17. John E. Young, *Mining the Earth*, Worldwatch Paper 109 (Washington, DC: Worldwatch Institute, July 1992); gold-to-waste ratio based on U.S. Bureau of Mines data provided in John E. Young, "Gold Production at Record High," in Lester R. Brown, Hal Kane, and David Malin Roodman, *Vital Signs 1994* (New York: W.W. Norton & Company, 1994); frontier forests from Bryant, Nielsen, and Tangley, op. cit. note 6.

18. Indigenous peoples figure from Roger Moody, "The Lure of Gold—How Golden Is the Future?" Panos Media Briefing No. 19 (London: Panos Institute, May 1996).

19. Minerals imports and exports are volume figures from UNCTAD, *Handbook of World Mineral Trade Statistics, 1992–1997* (Geneva: 1995).

20. U.S. mining industry from Christine A. Adamec, "Face Forward," *Mining Voice*, May/June 1996; 70 countries from William C. Symonds, "All that Glitters is Not Bre-X," *Business Week*, 19 May 1997.

21. Oil and gas reserves from BP Amoco, *BP Amoco Statistical Review of World Energy 1999* (London: 1999); *Oil and Gas Journal*, various issues.

22. Forest Stewardship Council (FSC) from FSC, "Who We Are," information sheet, <www.fscoax.org>, viewed 29 June 1999; growing demand from Sheila Polson, "'Green' Consumerism Starts to Benefit Some Forests," *Christian Science Monitor*, 31 March 1999; 17 million, 70 percent, and 30 countries from FSC, "Forests

Certified by FSC-Accredited Certification Bodies," information sheet, <www.fscoax.org/principal.htm>, viewed 11 October 1999; 1 million from "Limited Demand Seen for Goods Certified From Sustainable Forests, U.N. Report Says," *International Environment Reporter*, 28 October 1998; 200 million from World Wildlife Fund and World Bank, "What is the Forest Alliance?" information sheet, <www-esd.worldbank.org/wwf>, viewed 23 November 1999.

23. Trade in nontimber forest products from FAO, *State of the World's Forests 1997* (Rome: 1997); extractive reserves from John C. Ryan, "Goods From the Woods," *World Watch*, July/August 1991, and from Geoff Dyer, "Root and Branch Survival," *Financial Times*, 16–17 October 1999.

24. Crist Inman, "Impacts on Developing Countries of Changing Production and Consumption Patterns in Developed Countries: The Case of Ecotourism in Costa Rica," draft paper prepared for U.N. Environment Programme in conjunction with the Institute for Environmental Studies, Vrije Universiteit, Amsterdam, for "Global Product Chains: Northern Consumers, Southern Producers, and Sustainability," electronic conference hosted by the International Institute for Sustainable Development, <iisd.ca/susprod/ecotour.pdf>, viewed 5 December 1999; tourist arrivals from Rosa Songel, Department of Statistics and Economic Measurement of Tourism, World Tourism Organization, e-mail to Lisa Mastny, Worldwatch Institute, 25 November 1999; nature tourism figures from Fern L. Fillion, James P. Foley, and Andre J. Jaquemot, "The Economics of Global Ecotourism," paper presented at the Fourth World Congress on National Parks and Protected Areas, Caracas, Venezuela, 10–21 February 1992, as cited in The Ecotourism Society, "TES Statistical Fact Sheet," <www.ecotourism.org/tocfr.html>, viewed 14 January 1999, and from World Tourism Organization (WTO), *Yearbook of Tourism Statistics, Vol. 1* (Madrid: 1997); $1.6 billion and quote from WTO, "WTO Picks Hot Tourism Trends for 21st Century," press release (Madrid: 4 June 1998).

25. Environmental impacts of tourism from Dilys Roe, Nigel Leader-Williams, and Barry Dalal-Clayton, *Take Only Photographs, Leave Only Footprints*, IIED Wildlife and Development Series No. 10 (London: International Institute for Environment and Devel-

opment, October 1997); cultural impacts from Raymond Chavez, "Globalisation and Tourism: Deadly Mix for Indigenous Peoples," *Third World Resurgence*, March 1999; whale watching from Tim Padgett and Sharon Begley, "Beware of the Humans," *Newsweek*, 4 March 1996; Marcus Colchester and Fiona Watson, "Impacts of Ecotourism in Venezuela," November 1995, <www.txinfinet.com/mader/planeta/1195/1195ven.html>, viewed 28 April 1999.

26. For an example of possible local benefits from eco-tourism, see Erin Trowbridge, "Eco-tourism Profits Find Way Back to Mexican Communities," *The Earth Times*, 4 October 1998; Costa Rica from Inman, op. cit. note 24, and from Costa Rican Tourism Institute, "Certification for Sustainable Tourism," <www.turismo-sostenible.co.cr/EN/home.shtml>, viewed 12 October 1999; Zimbabwe from Cheri Sugal, "The Price of Habitat," *World Watch*, May/June 1997; Rwanda from Rainforest Action Network, "Can Ecotourism Save the Rainforests?" information sheet, <www.ran.org/ran/info_center>, viewed 9 December 1998.

27. Background on bioprospecting from Walter V. Reid et al., *Biodiversity Prospecting: Using Genetic Resources for Sustainable Development* (Washington, DC: WRI, May 1993), and from Colin Macilwain, "When Rhetoric Hits Reality in Debate on Bioprospecting," *Nature*, 9 April 1998; CBD Clearing-House Mechanism, op. cit. note 5.

28. Reid et al., op. cit. note 27; Jeffrey A. McNeely, "Achieving Financial Sustainability in Biodiversity Conservation Programs," in *Investing in Biodiversity Conservation, Workshop Proceedings* (Washington, DC: Inter-American Development Bank, July 1997).

29. Macilwain, op. cit. note 27.

30. Suriname from Layla Hughes, Conservation International, letter to Payal Sampat, Worldwatch Institute, 1 October 1997. For descriptions of other bioprospecting programs, see, for example, "Access to Genetic Resources: An Evaluation of the Development and Implementation of Recent Regulation and Access Agreements," Environmental Policy Studies Working Paper #4, prepared for the Biodiversity Action Network by the Environmental Policy Studies Workshop, 1999 (New York: Columbia University School of International and Public Affairs, June 1999), and

Julie M. Feinsilver, "Biodiversity Prospecting: A New Panacea for Development?" *Cepal Review*, December 1996.

31. Michael Totten, *Getting It Right: Emerging Markets for Storing Carbon in Forests* (Washington, DC: Forest Trends and WRI, 1999).

32. Ibid.

33. Ashley Mattoon, "Bogging Down In The Sinks," *World Watch*, November/December 1998; German Advisory Council on Global Change, *The Accounting of Biological Sinks and Sources Under the Kyoto Protocol: A Step Forwards or Backwards for Global Environmental Protection* (Bremerhaven, Germany: 1998).

CHAPTER 3. THE BIOTIC MIXING BOWL

1. Figure 3–1 based on U.N. Conference on Trade and Development (UNCTAD), *Review of Maritime Transport* (Geneva: various years), and on UNCTAD, "World Seaborne Trade Slows Down," press release (Geneva: 22 February 1999); Figure 3–2 based on International Civil Aviation Organization (ICAO), cited in Lisa Mastny, "Air Traffic Soaring," in Lester R. Brown, Michael Renner, and Brian Halweil, *Vital Signs 1999* (New York: W.W. Norton & Company, 1999), with 1998 figure from ICAO, "World Airline Passenger Traffic Growth to Pick Up Through to 2001," press release (Montreal: 21 July 1999).

2. This theme is fully set forth in a pioneering book by my colleague Chris Bright, which I have drawn on for this chapter; see Chris Bright, *Life Out of Bounds* (New York: W.W. Norton & Company, 1998), and Christopher Bright, "Invasive Species: Pathogens of Globalization," *Foreign Policy*, Fall 1999.

3. Bright, *Life Out of Bounds*, op. cit. note 2; world's vertebrates from World Resources Institute (WRI), *World Resources 1998–99* (New York: Oxford University Press, 1999); "almost half" from David S. Wilcove et al., "Quantifying Threats to Imperiled Species in the United States," *BioScience*, August 1998.

4. Bright, *Life Out of Bounds*, op. cit. note 2; John Yaukey, "Ships Carry Life, Chaos to Foreign Waters," *Gannett News Service*, 30 December 1998.

5. Bright, *Life Out of Bounds*.

6. Ibid.

7. Ibid.; WRI, op. cit. note 3; Lyle Glowka and Cyrille de Klemm, "International Instruments, Processes, Organizations and Non-Indigenous Species Introductions: Is a Protocol to the Convention on Biological Diversity Necessary?" in Odd Terje Sandlund, Peter Johan Schei, and Aslaug Viken, eds., *Proceedings of the Norway/UN Conference in Alien Species, Trondheim, 1–5 July 1996* (Trondheim, Norway: Directorate for Nature Management and Norwegian Institute for Nature Research, 1996).

8. Joby Warrick, "In U.S., An Asian Beetle Instills Full-Bore Economic Fear," *Washington Post*, 19 October 1998; Tom Baldwin, "Beetle Threat Could Force Change in Import Packaging," *Journal of Commerce*, 17 September 1998; European and Chinese restrictions from "China to Tighten Quarantine Rules on Wood Packing," *Reuters*, 2 November 1999.

9. World Wildlife Fund (WWF-US), "WWF Unveils Fourth 10 'Most Wanted' List," press release (Washington, DC: 3 June 1997).

10. Estimate of $10 million does not include timber and fisheries trade. This and species trade figures from TRAFFIC North America, "World Trade in Wildlife," information sheet (Washington, DC: July 1994); leading consumers from Peter H. Sand, "Commodity or Taboo? International Regulation of Trade in Endangered Species," in Helge Ole Bergesen and Georg Parmann, eds., *Green Globe Yearbook 1997* (Oxford: Oxford University Press, for the Fridtjof Nansen Institute, 1997); China and Southeast Asia from "Bear Parts Still in Demand in Asia," *Associated Press*, 26 October 1999, from Wendy Williams, "Turtle Tragedy," *Scientific American*, June 1999, and from Amy E. Vulpio, "From the Forests of Asia to the Pharmacies of New York City: Searching for a Safe Haven for Rhinos and Tigers," *Georgetown International Environmental Law Review*, vol. 11, no. 463 (1999); higher estimate of $20 million, illegal trade, and primate numbers from Statement by Shafqat Kakakhel, Deputy Executive Director, U.N. Environment Programme (UNEP), on behalf of the Executive Director, at UNEP Workshop on Enforcement and Compliance with Multilateral Environmental Agreements, Geneva, 12 July 1999.

11. Total of 146 countries from Convention on International Trade in Endangered Species of Wild Fauna and Flora (CITES)

Secretariat, "What is CITES?" <www.cites.org/CITES>, viewed 2 December 1999; species bans and trade restrictions from TRAFFIC North America, "CITES," information sheet (Washington, DC: October 1997); banned and restricted totals from CITES Secretariat, "Protected Species," <www.cites.org/CITES>, viewed 18 November 1999; CITES successes from Sand, op. cit. note 10, and from Organisation for Economic Co-operation and Development, "Experience With the Use of Trade Measures in the Convention on International Trade in Endangered Species of Wild Fauna and Flora (CITES)" (Paris: 1997).

12. History and effects of ivory ban from Caroline Taylor, "The Challenge of African Elephant Conservation," *Conservation Issues* (WWF-US), April 1997, and from TRAFFIC North America, "Elephant Ivory Trade," information sheet (Washington, DC: January 1997); recovery from "Kenyan Elephant Herd Recovering Slowly," *Environmental News Network*, <www.enn.com/news/wire-stories/1999/02/022299/elephants.asp>, 22 February 1999.

13. One-time trade from TRAFFIC International, "Ivory Trade Decisions at the 41st Meeting of the CITES Standing Committee," press release (Cambridge, U.K.: 11 February 1999); Kenya from "Call For Total Ban of Ivory Trade," *Panafrican News Agency*, 22 November 1999; Zimbabwe from Christopher Munnion, "Rangers Fear Rise in Ivory Poaching," *London Telegraph*, 24 November 1999.

14. "Half of World's Turtles Face Extinction, Scientists Say," *Environmental News Network*, <www.enn.com/news/enn-stories/1999/08/082699/freshturtle_5269.asp>, 26 August 1999; Williams, op. cit. note 10; TRAFFIC North America, "United States Supplies the World With Turtles," *Traffic North America Newsletter*, September 1998.

15. TRAFFIC North America, "Explosive Growth in US Live Reptile Trade Raises Conservation Fears," press release (Washington, DC: 10 September 1998); $30,000 from Michael Grunwald, "U.S. Bags Alleged Trafficker In Reptiles," *Washington Post*, 16 September 1998.

16. TRAFFIC International, "Yemeni Demand For Rhino Horn Daggers Continues," press release (Cambridge, U.K.: 7 May 1998).

17. On the role of the pet trade in bioinvasion, see Bright, *Life Out of Bounds*, op. cit. note 2; red-eared slider from TRAFFIC North America, op. cit. note 14.

18. Clive Ponting, *A Green History of the World* (New York: Penguin Books, 1991).

19. Mann quoted in Laurie Garrett, *The Coming Plague* (New York: Farrar, Straus, and Giroux, 1994).

20. Institute of Medicine, *America's Vital Interest in Global Health* (Washington, DC: National Academy Press, 1997).

21. World Health Organization (WHO), *Health and Environment in Sustainable Development: Five Years after the Earth Summit* (Geneva: 1997); David Pimentel et al., "Ecology of Increasing Disease," *BioScience*, October 1998.

22. HIV figures from Joint U.N. Programme on HIV/AIDS (UNAIDS), *AIDS Epidemic Update: December 1999* (Geneva: December 1999); Africa from UNAIDS, *Report on the Global HIV/AIDS Epidemic* (Geneva: June 1998).

23. Garrett, op. cit. note 19.

24. AIDS origins from "How AIDS Began," *The Economist*, 7 February 1998, and from Feng Gao et al., "Origin of HIV-1 in the Chimpanzee *Pan troglodytes troglodytes*," *Nature*, 4 February 1999; AIDS crossover from Donald G. McNeil, Jr., "The Great Ape Massacre," *New York Times Magazine*, 9 May 1999; human incursion from Jaap Goudsmit, "The Real Cause of the AIDS/HIV Epidemic: Destruction of Monkey and Ape Habitats in the African Rainforest," scientific abstract for presentation at the American Museum of Natural History's Spring Symposium on The Value of Plants, Animals, and Microbes to Human Health, 17–18 April 1998.

25. Bushmeat from Anthony L. Rose, The Bushmeat Project, "Growing Illegal Commerce in African Bushmeat Destroys Great Apes and Threatens Humanity," prepared for the American Zoo Association, 1999, <biosynergy.org/bushmeat/bmcommerce199.htm>, viewed 23 June 1999; Hahn quoted in McNeil, op. cit. note 24.

26. Garrett, op. cit. note 19.

27. Mann quote from ibid.

28. New diseases from WHO, *Report on Infectious Diseases: Removing Obstacles to Healthy Development* (Geneva: 1999); Lynne

Duke, "A Watch is Posted for Dead Crows," *Washington Post*, 29 September 1999; Andrew C. Revkin, "Mosquito Virus Exposes the Hole in the Safety Net," *New York Times*, 4 October 1999.

29. Quote and environmental changes from WHO, op. cit. note 21.

30. Rohit Burman, Kelly Kirschner, and Elissa McCarter, "Infectious Disease as a Global Security Threat," *Environmental Change and Security Project Report* (Woodrow Wilson Center), spring 1997; Peru and Americas infection and mortality figures from Pan American Health Organization, *Cholera Situation in the Americas* (Washington, DC: updated 24 July 1997); figure of $770 million from WHO, op. cit. note 28.

31. WHO, op. cit. note 28; WRI, op. cit. note 3.

32. Climate change from Paul R. Epstein, *Global Warming: Health and Disease* (Washington, DC: WWF-US, 1999); Epstein quoted in WWF-US, "Health Effects of Warming Could Be Devastating, WWF Report Finds," press release (Washington, DC: 5 November 1998).

33. Epstein, op. cit. note 32; "Extreme Weather's Effect on Health Measured," *Environmental News Network*, <www.enn.com/news/enn-stories/1999/02/021699/health.asp>, 16 February 1999.

34. History of international health cooperation from Richard N. Cooper, "International Cooperation in Public Health as a Prologue to Macroeconomic Cooperation," in Richard N. Cooper et al., *Can Nations Agree?* (Washington, DC: The Brookings Institution, 1989).

CHAPTER 4. GLOBAL GROCERS

1. Chiapas uprising from Michael Renner, "Chiapas: The Fruits of Despair," *World Watch*, January/February 1997; effects of North American Free Trade Agreement (NAFTA) on farmers from U.N. Development Programme (UNDP), *Human Development Report 1997* (New York: Oxford University Press, 1997).

2. Importance of agriculture from Walden Bello, "Asia, Asian Farmers, and the WTO," *Focus on the Global South*, July 1999; figures on agriculture's importance in low-income countries from "Further Liberalization in Agriculture Will Be Pushed By Develop-

ing Countries; Other Ag Trade Issues," *BRIDGES Weekly Trade New Digest*, 3 May 1999, and from U.N. Food and Agriculture Organization (FAO), *FAOSTAT Statistics Database*, electronic database, <apps.fao.org>, viewed 5 December 1999; Japanese negotiator quoted in Elizabeth Olson, "Serious Issues Behind Divisions as WTO Searches for a Leader," *New York Times*, 15 May 1999.

3. Figure 4–1 from FAO, op. cit. note 2, viewed 17 November 1999, with figures deflated using U.S. GNP Implicit Price Deflator from U.S. Department of Commerce, *Survey of Current Business*, July 1999; 11 percent and continental breakdown from World Trade Organization (WTO), *Annual Report 1998* (Geneva: 1998); relative importance of food grains versus nonessentials from FAO, op. cit. note 2, viewed 3 December 1999.

4. Grain trade statistics from U.S. Department of Agriculture (USDA), *Production, Supply, and Distribution*, electronic database, Washington, DC, updated 9 April 1999; FAO low-income food-deficit countries from FAO, Special Programme for Food Security, "Low-income Food-deficit Countries," information sheet, <www.fao.org/spfs/lifdc-e.htm>, viewed 22 November 1999.

5. Grain trade during Roman empire from Susan Raven, *Rome in Africa* (New York: Routledge, 1993); historical trends from Lester R. Brown, *Who Will Feed China?* (New York: W.W. Norton & Company, 1995), derived and updated from USDA, op. cit. note 4, November 1994 and December 1999.

6. Brown, op. cit. note 5.

7. FAO, op. cit. note 2, viewed 6 December 1999.

8. Developing countries' share of exports (by volume) from ibid.; long-term decline in commodity prices from World Bank, *Global Development Finance 1999* (Washington, DC: 1999); Central and South America from Lori Ann Thrupp, *Bittersweet Harvests for Global Supermarkets* (Washington, DC: World Resources Institute (WRI), 1995); Chile from Nick Robins and Sarah Roberts, *Unlocking Trade Opportunities: Case Studies of Export Success from Developing Countries*, prepared for the U.N. Department of Policy Co-ordination and Sustainable Development (London: International Institute for Environment and Development, 1998).

9. Angela Paxton, *The Food Miles Report: The Dangers of Long-Distance Food Transport* (London: National Food Alliance, 1994);

Wuppertal Institute, *Road Transport of Goods and the Effects on the Spatial Environment*, July 1993, cited in Richard Douthwaite, *Short Circuit* (Devon, U.K.: Green Books, 1996).

10. Sandra Postel, *Pillar of Sand* (New York: W.W. Norton & Company, 1999).

11. Ibid.

12. For a discussion of proposals for international water transport, see Peter H. Gleick, *The World's Water: The Biennial Report on Freshwater Resources 1998–1999* (Washington, DC: Island Press, 1998); Canadian case from Anthony DePalma, "Free Trade in Fresh Water? Canada Says No," *New York Times*, 7 March 1999; International Joint Commission, "IJC to Hold Public Hearings on Uses, Diversions, and Bulk Exports of Great Lakes Water," press release (Washington, DC: 22 February 1999).

13. On the implications of NAFTA and the WTO, see Maude Barlow, *Blue Gold: The Global Water Crisis and the Commodification of the World's Water Supply* (San Francisco, CA: International Forum on Globalization, June 1999); on the Sun Belt case, see "Canada's Efforts to Protect Water Hit Turbulence," *Reuters*, 2 December 1999.

14. For implications of agricultural transformations on women, see, for example, Jodi Jacobson, *Gender Bias: Roadblock to Sustainable Development*, Worldwatch Paper 110 (Washington, DC: Worldwatch Institute, September 1992); Vandana Shiva, *Staying Alive: Women, Ecology and Survival in India* (New York: St. Martin's Press, 1989); and FAO, "Women Feed the World," prepared for World Food Day, 16 October 1998 (Rome: 1998). For a discussion of the social fallout of export agriculture in Costa Rica, see Alicia Korten, *A Bitter Pill: Structural Adjustment in Costa Rica* (Oakland, CA: Institute for Food and Development Policy, June 1995).

15. Recent fires from "Indonesian Fires Blamed on Plantations," *Reuters*, 10 August 1999, from "With Nothing Left to Burn, Fires Mostly Out," *Washington Post*, 25 April 1998, and from "Asian Nations Reach Accord on Fighting Haze," *Washington Post*, 24 December 1997; palm oil exports from FAO, op. cit. note 2, viewed 6 December 1999; effects of International Monetary Fund program on Indonesia's forests from William D. Sunderlin, Con-

sultative Group on International Agricultural Research, "Between Danger and Opportunity: Indonesia's Forests in an Era of Economic Crisis and Political Challenge" (Bogor, Indonesia: Center for International Forestry Research, 11 September 1998); Amazon Basin from Atossa Soltani and Tracey Osborne, *Arteries for Global Trade, Consequences for Amazonia* (Malibu, CA: Amazon Watch, April 1997), and from Peter May, Ana Célia Castro, and Antonio Barros de Castro, "Expansion and Technical Innovation in Brazil's Soybean-Based Agroindustrial Complex," in Bradford S. Gentry, ed., *Private Capital Flows and the Environment: Lessons from Latin America* (Cheltenham, U.K.: Edward Elgar Press, 1998).

16. Colombian study and 20 percent reported in Thrupp, op. cit. note 8; Colombia Human Rights Committee study cited in Pesticide Action Network North America, "Colombian Flower Worker Dismissed from Job Following U.S. Tour," press release (San Francisco, CA: 26 May 1993); U.S. market from Larry Rohter, "Foreign Presence in Colombia's Flower Gardens," *New York Times*, 8 May 1999.

17. Thrupp, op. cit. note 8.

18. Central American beef from Cary Fowler and Pat Mooney, *Shattering: Food, Politics and the Loss of Genetic Diversity* (Tucson: University of Arizona Press, 1990); U.S. appetite from Alan B. Durning, "Fat of the Land," *World Watch*, May/June 1991; Botswana land degradation from Alan B. Durning and Holly B. Brough, *Taking Stock: Animal Farming and the Environment*, Worldwatch Paper 103 (Washington, DC: Worldwatch Institute, July 1991); Botswana beef exports from FAO, op. cit. note 2, viewed 6 December 1999; Somalia from Bruce Byers, "Roots of Somalia's Crisis," *Christian Science Monitor*, 24 December 1992.

19. Economic dependence on fish from Anne Platt McGinn, *Rocking the Boat: Conserving Fisheries and Protecting Jobs*, Worldwatch Paper 142 (Washington, DC: June 1998); share of animal protein from Edmondo Laureti, *Fish and Fishery Products: World Apparent Consumption Statistics Based on Food Balance Sheets (1961–1995)* (Rome: FAO, November 1998); 1 billion and Asian dependence from WRI, *World Resources 1996–97* (New York: Oxford University Press, 1996); fishing grounds from Maurizio Perotti, Fishery Data, Information, and Statistics Unit, FAO Fish-

eries Department, e-mail to Anne Platt McGinn, Worldwatch Institute, 14 October 1997; overexploitation from FAO, *The State of World Fisheries and Aquaculture, 1996* (Rome: 1997).

20. Data for Figure 4–2 and share of industrial and developing countries from Sara Montanaro, Statistical Clerk, Fishery Information, Data, and Statistics Unit, FAO Fisheries Department, Rome, e-mail to Lisa Mastny, 6 September 1999, and from FAO, *Yearbook of Fishery Statistics*, vol. 45 (Rome: 1978); share of fish traded today (1997 data) from FAO, *Yearbook of Fishery Statistics*, vol. 84 (Rome: 1999), and from FAO, *Yearbook of Fishery Statistics*, vol. 85 (Rome: 1999); share traded in 1970 from FAO, vol. 45, op. cit. this note; four developing countries from FAO, *Yearbook of Fishery Statistics*, vol. 53 (Rome: 1983), and from FAO, vol. 85, op. cit. this note. All figures expressed in 1998 dollars and deflated as described in note 3.

21. Law of the Sea from United Nations, Division for Ocean Affairs and the Law of the Sea, "United Nations Convention on the Law of the Sea," <www.un.org/Depts/los/unclos/closindx.htm>, viewed 23 November 1999; long-distance fleets from World Wide Fund for Nature (WWF International), *The Footprint of Distant Water Fleets on World Fisheries* (Gland, Switzerland: 1998); African access agreements from Gareth Porter, "Euro-African Fishing Agreements: Subsidizing Overfishing in African Waters," in Scott Burns, ed., *Subsidies and Depletion of World Fisheries: Case Studies* (Washington, DC: Endangered Seas Campaign, World Wildlife Fund, April 1997), and from Okechukwu C. Iheduru, "The Political Economy of Euro-African Fishing Agreements," *The Journal of Developing Areas*, October 1995; Pacific access from Lisa K. Bostwick, "Empowering South Pacific Fishmongers: A New Framework for Preferential Access Agreements in the South Pacific Tuna Industry," *Law & Policy in International Business*, vol. 26 (1995).

22. Deprivation of small-scale fishers from George Kent, "Fisheries, Food Security, and the Poor," *Food Policy*, vol. 22, no. 5 (1997), from George Kent, *Nutrition Rights and Fisheries*, prepared for Greenpeace International, 19 September 1994, and from Mahfuzuddin Ahmed, "Fish for the Poor Under a Rising Global Demand and Changing Fishery Regime," *NAGA, The ICLARM Quarterly*, July/December 1997; Senegal from Kelly Haggart,

"Exporting More, Eating Less," *Panoscope*, November 1992, and from Harry de Vries, "The Fight for Fish: Towards Fair Fisheries Agreements" (Brussels: Eurostep, May 1996).

23. Ratio of one to five from FAO, *State of World Fisheries and Aquaculture 1998* (Rome: 1999); other information from McGinn, op. cit. note 19.

24. UNDP, *Human Development Report 1998* (New York: Oxford University Press, 1998); Thai shrimp production from FAO, op. cit. note 2; Ecuadorean and Thai shrimp exports from FAO, vol. 85, op. cit. note 20; Thai shrimp aquaculture production from FAO, *Aquaculture Production Statistics 1988–1997* (Rome: 1999); effects of Asian crisis from "Thailand Sees Lower Shrimp Production in 1999," *Reuters*, 7 June 1999; Ecuadorean export earnings from International Monetary Fund, *International Financial Statistics* (Washington, DC: September 1999); repercussions from Mike Hagler, "Shrimp—The Devastating Delicacy: The Explosion of Shrimp Farming and the Negative Impacts on People and the Environment," *Ecological Economics Bulletin*, Third Quarter 1998.

25. Hagler, op. cit. note 24.

26. Varieties of sweet potato from John C. Ryan, *Life Support: Conserving Biodiversity*, Worldwatch Paper 108 (Washington, DC: Worldwatch Institute, April 1991). For a general discussion of the value of and threats to plant genetic diversity, see John Tuxill, *Nature's Cornucopia: Our Stake in Plant Diversity*, Worldwatch Paper 148 (Washington, DC: Worldwatch Institute, September 1999).

27. Loss of 75 percent of crop diversity and Mexico from FAO, *State of the World's Plant Genetic Resources* (Rome: 1996); role of Green Revolution from Tuxill, op. cit. note 26; U.S. corn from Ryan, op. cit. note 26; India from Reed Karaim, "Variety, the Vanishing Crop," *Washington Post*, 11 April 1999.

28. Clive Jones, *PREVIEW: Global Review of Commercialized Transgenic Crops: 1999* (Ithaca, NY: International Service for the Acquisition of Agri-Biotech Applications, 1999); Scott Kilman, "Once Quick Converts, Farmers Begin to Lose Faith in Biotech Crops," *Wall Street Journal*, 19 November 1999.

29. For various perspectives on biotechnology's role, see

Brian Halweil, "The Emperor's New Crop," *World Watch*, July/August 1999; FAO, "Biotechnology," Item 7 of the Provisional Agenda for the 15th Session of the FAO Committee on Agriculture, held in Rome, 25–29 January 1999, <www.fao.org/unfao/bodies/COAG/COAG15/X0074E.htm>, viewed 21 May 1999; and John Madeley, "Genetic Engineering Will Not Feed Hungry, Say Africans," *Financial Times*, 16 March 1999.

30. Halweil, op. cit. note 29; 80 percent from Steve Suppan, "Biotechnology's Takeover of the Seed Industry," information sheet (Minneapolis, MN: Institute for Agriculture and Trade Policy (IATP), 13 July 1998).

31. Controversial patents from Genetic Resources Action International, "Patenting Life: Progress or Piracy?" *Global Biodiversity*, spring 1998; consolidation and industry giants from Halweil, op. cit. note 29, and from "Private Parts: Privatisation and the Life Industry," *Development Dialogue*, Special Issue 1996.

32. Tuxill, op. cit. note 26; Kristin Dawkins and Steve Suppan, *Sterile Fields: The Impacts of Intellectual Property Rights and Trade on Biodiversity and Food Security* (Minneapolis, MN: IATP, November 1996); Geoff Tansey, "Trade, Intellectual Property, Food and Biodiversity," a discussion paper commissioned by Quaker Peace and Service, London, in association with Quaker United Nations Office, Geneva, February 1999.

33. For a description of the WTO's Agreement on Agriculture, see Jeffrey S. Thomas and Michael A. Meyer, *The New Rules of Global Trade* (Scarborough, ON, Canada: Carswell Thomas Professional Publishing, 1997).

34. Historical backdrop from WWF International, *Agriculture in the Uruguay Round: Implications for Sustainable Development in Developing Countries* (Gland, Switzerland: January 1995); Robert Repetto, *Trade and Sustainable Development*, UNEP Environment and Trade Series No. 1 (Nairobi: U.N. Environment Programme, 1994).

35. Caroline LeQuesne, *Reforming World Trade: The Social and Environmental Priorities* (Oxford: Oxfam Publications, 1996).

36. The least developed countries are exempt from the requirements to phase out import restrictions, per LeQuesne, op. cit. note 35, and Thomas and Meyer, op. cit. note 33; information

on agriculture agreement from these sources and from WWF International, op. cit. note 34; Laranjo quoted in Kevin Watkins, "Free Trade and Farm Fallacies: From the Uruguay Round to the World Food Summit," *The Ecologist*, November/December 1996.

37. Bello, op. cit. note 2. See also "Japan to Take Food Security Stance in WTO Ag Talks," *BRIDGES Weekly Trade News Digest*, 26 July 1999, and "Agricultural Subsidies Again Emerge as Top Priority of New WTO Round," *BRIDGES Between Trade and Sustainable Development*, April 1999.

38. On the benefits of local agriculture, see Richard Douthwaite, *Short Circuit: Strengthening Local Economies for Security in an Unstable World* (Foxhole, Dartington, U.K.: Green Books, 1996), and Molly O'Meara, *Reinventing Cities for People and the Planet*, Worldwatch Paper 147 (Washington, DC: Worldwatch Institute, June 1999).

39. Background on fair trade movement from Vicki Elkin, "Fairer Trade," *World Watch*, July/August 1992, and from Bert Beekman, "Fair Trade and Trade Development," *ILEIA Newsletter*, December 1998; Dutch coffee imports from Stichting Max Havelaar, "Essential Points from the 1998 Annual Report Max Havelaar Foundation," information sheet (Utrecht, Netherlands: 1999); TransFair from Robert A. Rice and Justin R. Ward, *Coffee, Conservation, & Commerce in the Western Hemisphere* (Washington, DC: Smithsonian Migratory Bird Center and Natural Resources Defense Council, June 1996).

40. Rice and Ward, op. cit. note 39; Alexandra Marks, "Environmentalists Target Java-Drinkers to Save Birds," *Christian Science Monitor*, 10 July 1997.

41. Value of world organic market from Bernward Geier, "The Organic Market: Opportunities and Challenges," *ILEIA Newsletter*, December 1998; projected growth from Peter Ford, "Organic Farmers Hear a Call: If You Grow It, They Will Buy," *Christian Science Monitor*, 24 March 1999; $4.7 billion from Organic Trade Association, "Organic Trade Association Questions Timing of USDA 'Issue Papers' on Organic Rule," press release (Greenfield, MA: 26 October 1998); 3–5 million consumers is a conservative estimate from Gary Gardner, "Organic Farming Up Sharply," in Lester R. Brown, Christopher Flavin, and Hal Kane, *Vital Signs*

1996 (New York: W.W. Norton & Company, 1996).

42. Gardner, op. cit. note 41; Robins and Roberts, op. cit. note 8.

43. Gary C. Groves, "Update on Argentina's Organic Sector 1998," Global Agriculture Information Network Report #AR8066 (Buenos Aires: USDA, Foreign Agricultural Service, 15 October 1998).

44. Andrew Wheat, "Toxic Bananas," *Multinational Monitor*, September 1996; Rainforest Alliance programs from Rainforest Alliance, "The Conservation Agriculture Program," and "Program Description," <www.rainforest-alliance.org/programs/cap>, viewed 13 October 1999; Chiquita Brands, Inc., "Better Banana Project," information sheet, <www.envirochiquita.com>, viewed 13 October 1999; "fair trade banana" from Andrea Spencer-Cooke, "Eco-Labeling Goes Bananas," *Tomorrow*, September/October 1997.

45. Marine Stewardship Council (MSC), "Our Empty Seas: A Global Problem, A Global Solution," information brochure (London: April 1999); membership from MSC, "MSC Signatories," <www.msc.org>, viewed 23 November 1999; Unilever pledge from MSC, "Position Paper Submitted to the Seventh Session of the UN Commission on Sustainable Development 1999 (CSD-7)," New York, 19 April 1999; MSC symbol debut from MSC, "First Certifications in Closing Stages," *MSC News*, October 1999.

46. WWF International, "Directing WTO Negotiations toward Sustainable Agriculture and Rural Development," WWF Discussion Paper (Gland, Switzerland: November 1999); Mark Musick, "Report on NGO Food and Agriculture Day in Seattle," on list serve <road_to_seattle@iatp.org>, posted 13 December 1999.

47. Joseph Kahn and David E. Sanger, "Trade Obstacles Unmoved, Seattle Talks End in Failure," *New York Times*, 4 December 1999.

CHAPTER 5. THE EXPORT OF HAZARD

1. Bhopal events and quote from Sanjoy Hazarika, *Bhopal: The Lessons of a Tragedy* (New Delhi: Penguin Books, 1987); deaths from Satinath Sarangi, "The Union Carbide Disaster In Bhopal—A Report From 1996" (Bhopal, India: Bhopal Group for

Information and Action, 1996).

2. Ash disposal attempts from Hilary F. French, "A Most Deadly Trade," *World Watch*, July/August 1990; eventual dumping from Greenpeace International, "The Basel Ban—The Pride of the Basel Convention" (Amsterdam: September 1995).

3. Nana quote from French, op. cit. note 2; Greenpeace International, op. cit. note 2; Center for Investigative Reporting (CIR) and Bill Moyers, *Global Dumping Ground: The International Traffic in Hazardous Waste* (Washington, DC: Seven Locks Press, 1990).

4. Waste import bans and history of convention from Greenpeace International, op. cit. note 2; 2.6 million tons from Greenpeace International, *The Database of Known Hazardous Waste Exports from OECD to Non-OECD Countries, 1989–March 1994* (Washington, DC: 1994), cited in World Resources Institute (WRI), *World Resources 1998–99* (New York: Oxford University Press, 1999); documented dumpings from Aaron Sachs, *Eco-Justice: Linking Human Rights and the Environment*, Worldwatch Paper 127 (Washington, DC: Worldwatch Institute, December 1995), citing U.N. Environment Programme (UNEP), "Generation of Hazardous Wastes and Other Wastes, 1993 Statistics," paper prepared for the Third Meeting of the Conference of the Parties to the Basel Convention, Geneva, 18–22 September 1995.

5. Unilateral bans and quote from Greenpeace International, op. cit. note 2; requirements for ratification from Camila Reed, "INTERPOL-UNEP Link-Up to Attack Toxic Waste Trade," *Reuters*, 24 February 1999; 13 countries from United Nations, *The United Nations Treaty Collection*, electronic database, <www.un.org/Depts/Treaty>, viewed 5 December 1999; opposition from Basel Action Network (BAN), "Basel Ban Victory at COP4," <www.ban.org>, viewed 5 June 1999.

6. UNEP, "Hazardous Wastes Talks to Consider Liability Regime for Basel Convention," press release (Nairobi: 9 April 1999); Portas quote from Reed, op. cit. note 5; illegal shipments from WRI, op. cit. note 4.

7. Rone Tempest, "Asia's Toxic Formula for Waste," *Los Angeles Times*, 4 March 1999; "WHO Says Waste Dumped in Cambodia May Pose Long-Term Threat to Health," *International Environment Reporter*, 6 January 1999.

8. Taiwanese awareness in Tempest, op. cit. note 7; Environmental Protection Agency reversal from BAN, "Victory for Environmental Justice: Mercury Waste from Taiwan Dumped in Cambodia Will Not Be Re-dumped in California," press release (Seattle, WA: 31 March 1999); other disposal efforts and recent events from BAN, "Asian Poisons to France: Green Groups Condemn Plan of 'Free Trade' In Toxic Waste," press release (Seattle, WA: 30 September 1999), and from Chiu Yu-Tzu, "France Says No to FPG's Toxic Waste," *Taipei Times*, 10 November 1999.

9. Fred Pearce, "Northern Exposure," *New Scientist*, 31 May 1997.

10. Theo Colburn, Dianne Dumanoski, and John Peterson Myers, *Our Stolen Future* (New York: Plume Books, March 1997); Pearce, op. cit. note 9.

11. Colburn, Dumanoski, and Myers, op. cit. note 10; Éric Dewailly et al., "Concentration of Organochlorines in Human Brain, Liver, and Adipose Tissue Autopsy Samples from Greenland," *Environmental Health Perspectives*, October 1999.

12. Pesticide use from David Pimental, Cornell University, e-mail to Brian Halweil, Worldwatch Institute, 15 January 1999; toxicity levels from David Pimentel et al., "Ecology of Increasing Diseases," *BioScience*, October 1998; Figure 5–1 based on data in U.N. Food and Agriculture Organization (FAO), *FAOSTAT Statistics Database*, electronic database, <www.apps. fao.org>, viewed 30 November 1999, with figures deflated using the U.S. G.N.P. Implicit Price Deflator provided in U.S. Department of Commerce, *Survey of Current Business*, July 1999; pesticide bans and "dirty dozen" from Pesticide Action Network North America, "Over 90 Countries Have Banned 'Dirty Dozen' Pesticides," press release (San Francisco: 23 October 1995); "circle of poison" from David Weir and Mark Schapiro, *Circle of Poison : Pesticides and People in a Hungry World* (Oakland, CA: Food First Books, February 1981).

13. Lack of gear and legible labels from Barbara Dinham, *The Pesticide Hazard* (London: Zed Books, 1993); vague labeling from Foundation for Advancements in Science and Education (FASE), *Exporting Risk: Pesticide Exports from US Ports 1995–96* (Los Angeles, CA: spring 1998); World Health Organization estimates of pesticide poisonings cited in Barbara Dinham, "The Success of a

Voluntary Code in Reducing Pesticide Hazards in Developing Countries," in Helge Ole Bergesen and Georg Parmann, eds, *Green Globe Yearbook 1996* (Oxford: Oxford University Press, for the Fridtjof Nansen Institute, 1996).

14. Jim Morris, "Exporting Sterility: Perils of the Pesticide Trade," *U.S. News and World Report*, 10 May 1999; Costa Rican imports from Andrew Wheat, "Toxic Bananas," *Multinational Monitor*, September 1996.

15. Class action suit and Levy quote from Morris, op. cit. note 14.

16. FASE, op. cit. note 13.

17. Ibid.

18. FAO, "FAO Warns of the Dangerous Legacy of Obsolete Pesticides," press release (Rome: 24 May 1999).

19. Ibid.

20. Industry-government collaboration from Jonathan Dahl, "Canada Encourages Mining of Asbestos, Sells to Third World," *Wall Street Journal*, 12 September 1989; Canadian exports from Bill Schiller, "Why Canada Pushes Killer Asbestos," *Toronto Star*, 20 March 1999; World Trade Organization (WTO) and European Union from Daniel Pruzin, "WTO Ruling on France's Asbestos Ban Delayed Until March 2000, EU Official Says," *International Environment Reporter*, 29 September 1999.

21. "Treaty on Prior Informed Consent Signed By 57 Countries at Rotterdam Meeting," *International Environment Reporter*, 16 September 1998; FAO Code of Conduct from Dinham, *The Pesticide Hazard*, op. cit. note 13; on negotiations on persistent organic pollutants, see Anne Platt McGinn, "Phasing Out Persistent Organic Pollutants," in Lester R. Brown et al., *State of the World 2000* (New York: W.W. Norton & Company, 2000).

22. Production trends and asbestos deaths from Dennis Cauchon, "The Asbestos Epidemic—A Global Crisis," *USA Today*, 8 February 1999; Brazil from WRI, op. cit. note 4.

23. Car battery exports and health concerns from CIR and Moyers, op. cit. note 3; Brazil, India, and China from Greenpeace International, *Lead Astray Again: The Ongoing Illegal Trade of U.S. Scrap Lead Acid Batteries to Brazil* (Amsterdam: August 1997).

24. Share of production in developing countries from U.N.

Industrial Development Organization, *International Yearbook of Industrial Statistics* (Vienna: 1999); U.S. foreign direct investment outflows from U.S. Department of Commerce, *Survey of Current Business*, September 1999.

25. High-tech hazards from Bill Richards, "Semiconductor Plants Aren't Safe and Clean As Billed, Some Say," *Wall Street Journal*, 5 October 1998, and from John E. Young, *Global Network: Computers in a Sustainable Society,* Worldwatch Paper 115 (Washington, DC: Worldwatch Institute, September 1993); 29 sites from Ted Smith, "The Dark Side of High-Tech Development" (San Jose, CA: Silicon Valley Toxics Coalition (SVTC), 11 August 1998).

26. Spread of high-tech and Philippines figures from Chito Salazar, "Semi-Conductors from the Philippines," paper prepared for UNEP and for "Global Product Chains: Northern Consumers, Southern Producers, and Sustainability," electronic conference, hosted by the International Institute for Sustainable Development, <iisd.ca/susprod/semiconduct.pdf>, viewed 14 June 1999; SVTC review from Carlos Plazola, "The Globalization of High Tech: Environmental Injustices Plague Industry" (San Jose, CA: SVTC, 1996), and from SVTC, "Global Manufacturing and Assembly Facilities for the Computer Manufacturing Industry," information sheet (San Jose, CA: April 1997).

27. For a review of the literature on this subject, see WTO, *Trade and the Environment* (Geneva: 1999).

28. For a full though somewhat dated discussion of the role of environmental factors in companies' location decisions, see H. Jeffrey Leonard, *Pollution and the Struggle for the World Product* (Cambridge, U.K.: Cambridge University Press, 1988); on the North American Free Trade Agreement (NAFTA) debate's focus on the border area, see, for example, John Holusha, "Trade Pact May Intensify Problems at the Border," *New York Times*, 20 August 1992; number of manufacturing plants from International Labour Organisation (ILO), "Labour and Social Issues Relating to Export Processing Zones" (Geneva: August 1998); Mexicali from Roberto Sanchez, "Health and Environmental Risks of the Maquiladora in Mexico," *Natural Resources Journal*, Winter 1990.

29. S.J. Lewis et al., *Border Trouble: Rivers in Peril* (Boston: National Toxics Campaign Fund, 1991). On the lack of progress

since NAFTA, see Leslie Crawford, "Hazardous Trades Bring Pollution and Health Fears Down Mexico Way," *Financial Times*, 6
June 1997; Public Citizen, *NAFTA's Broken Promises: The Border
Betrayed* (Washington, DC: 1997); and Public Citizen, *NAFTA at
Five Years Report Card* (Washington, DC: 22 December 1998). For
a more optimistic view, see Commission for Environmental Coop-
eration, *Assessing Environmental Effects of the North American Free
Trade Agreement* (Montreal: 1999). See also "A Greener, or Brown-
er, Mexico?" *The Economist*, 7 August 1999. For a discussion of
ongoing institutional initiatives, see Office of the United States
Trade Representative, *Study on the Operation and Effects of the
NAFTA* (Washington, DC: July 1997).

30. Export processing zone figures from ILO, op. cit. note 28;
inducements and attitudes from Alexander Goldsmith, "Seeds of
Exploitation," in Jerry Mander and Edward Goldsmith, eds., *The
Case Against the Global Economy* (San Francisco, CA: Sierra Club
Books, 1996).

31. Cavite information from Keith B. Richburg, "Under
Southeast Asia's Haze: More Bad Air," *Washington Post*, 5 October
1997; Chinese accusation from Xia Guang, "Pollution by Foreign
Firms Rises," *China Environment News*, 15 February 1997.

32. For positive links between foreign investment and environmental performance, see WTO, op. cit. note 27; Gunnar S.
Eskeland and Ann E. Harrison, *Moving to Greener Pastures? Multinationals and the Pollution-Haven Hypothesis* (Washington, DC:
World Bank, March 1997); and Global Environmental Management Initiative, *Fostering Environmental Prosperity: Multinationals
in Developing Countries* (Washington, DC: February 1999). Yonghai quote from Zhang Yan, "Traders Pay Attention to Environment," *China Daily*, 12 April 1999.

33. On the effect of privatization on the environment, see
Pamela Stedman-Edwards et al., *The Private Sector in Latin America*
(Washington, DC: World Wildlife Fund, July 1997); Bradford S.
Gentry, *Private Investment and the Environment*, Discussion Paper 11
(New York: U.N. Development Programme, undated); and Bradford
S. Gentry and Lisa Fernandez, "Mexican Steel," in Bradford S. Gen-
try, ed., *Private Capital Flows and the Environment: Lessons from Latin
America* (Cheltenham, U.K.: Edward Elgar Publications, 1999).

CHAPTER 6. SHARING THE AIR

1. J.C. Farman, B.G. Gardiner, and J.D. Shanklin, "Large Losses of Total Ozone in Antarctica Reveal Seasonal ClO_x/NO_x Interaction," *Nature*, 16 May 1985; Douglas C. Cogan, *Stones in a Glass House* (Washington, DC: Investor Responsibility Research Center, 1988).

2. For a description of the early industry intransigence, see Paul Brodeur, "The Annals of Chemistry," *The New Yorker*, 9 June 1986.

3. The impact of the ozone hole findings on the treaty negotiations is a matter of some debate. On this subject, see Richard Elliot Benedick, *Ozone Diplomacy*, enlarged (Cambridge, MA: Harvard University Press, 1998); Brodeur, op. cit. note 2; and Peter Morrisette, "The Evolution of Policy Responses to Stratospheric Ozone Depletion," *Natural Resources Journal*, vol. 29 (1989). On the positioning of industry, see Benedick, op. cit. this note; Ian H. Rowlands, *The Politics of Global Atmospheric Change* (Manchester, U.K.: Manchester University Press, 1995); and Forest Reinhardt, "DuPont Freon Products Division (A)," Harvard Business School case study (Cambridge, MA: 28 March 1995).

4. Ozone Secretariat, *The 1987 Montreal Protocol on Substances That Deplete the Ozone Layer*, <www.unep.ch/ozone/mont _t.htm>, viewed 2 December 1999.

5. International Institute for Sustainable Development (IISD), "Report of the Third Conference of the Parties to the United Nations Framework Convention on Climate Change: 1–11 December 1997," *Earth Negotiations Bulletin*, 13 December 1997.

6. Benedick, op. cit. note 3; ratifications from United Nations, *The United Nations Treaty Collection*, electronic database, <www.un.org/Depts/Treaty>, viewed 6 December 1999.

7. Phaseout in industrial countries was with the exception of a few "essential uses," and of production of limited quantities in order to meet the basic domestic needs of developing countries; Elizabeth Cook, "Marking a Milestone in Ozone Protection: Learning from the CFC PhaseOut" (Washington, DC: World Resources Institute, January 1996). Figure 6–1 compiled by Worldwatch Institute from the following sources: 1950 and 1955,

estimates based on Chemical Manufacturers Association; 1960–95 from Sharon Getamel, DuPont, Wilmington, DE, letter to World-watch, 15 February 1996; and 1996–97 from U.N. Environment Programme (UNEP), *Data Report on Production and Consumption of Ozone Depleting Substances, 1986–1998* (Nairobi: October 1999). Data are measured in tons weighted by ozone depletion potential.

8. UNEP, op. cit. note 7; record ozone holes from World Meteorological Organization (WMO), "Record Ozone Depletion in the Antarctic," press release (Geneva: 1 October 1998), and from UNEP, "Antarctic Ozone Hole as Big as Ever," press release (Nairobi: 17 November 1999).

9. Ozone Secretariat, *Synthesis of the Reports of the Scientific, Environmental Effects, and Technology and Economic Assessment Panels of the Montreal Protocol: A Decade of Assessments for Decision Makers Regarding Protection of the Ozone Layer: 1988–1999* (Nairobi: UNEP, February 1999).

10. Seth Dunn, "Carbon Emissions Dip," in Lester R. Brown, Michael Renner, and Brian Halweil, *Vital Signs 1999* (New York: W.W. Norton & Company, 1999); temperature projections from J.T. Houghton et al., eds., *Climate Change 1995: The Science of Climate Change*, Contribution of Working Group I to the Second Assessment Report of the Intergovernmental Panel on Climate Change (IPCC) (Cambridge, U.K.: Cambridge University Press, 1996).

11. Projection of 30 percent from Michael Grubb and Christiaan Vrolijk, "The Kyoto Protocol: Specific Commitments and Flexibility Mechanisms," *Climate Change Briefing No. 11* (London: Royal Institute of International Affairs, April 1998), based on International Energy Agency projections; cuts of 60–80 percent from Houghton et al., op. cit. note 10.

12. IPCC quote from Houghton et al., op. cit. note 10; Figure 6–2 from James Hansen et al., Goddard Institute for Space Studies, Surface Air Temperature Analysis, "Global Temperature Anomalies in .01 C, 1866–1998," as posted at <www.giss.nasa.gov/data/update/gistemp>, viewed 6 December 1999.

13. Robert T. Watson, Marufu C. Zinyowera, and Richard H. Moss, eds. *Climate Change 1995: Impacts, Adaptation, and Mitigation*, Contribution of Working Group II to the Second Assessment

Report of the IPCC (Cambridge, U.K.: Cambridge University Press, 1996); Molly O'Meara, "The Risks of Disrupting Climate," *World Watch*, November/December 1997; Hadley Centre for Climate Prediction and Research, *Climate Change and Its Impacts: Some Highlights from the Ongoing UK Research Programme: A First Look at Results from the Hadleys Centre's New Climate Model* (Bracknell, U.K.: November 1998).

14. United Nations, "United Nations Framework Convention on Climate Change, 1992," <www.unfccc.org>, viewed 2 December 1999.

15. Vienna Convention for the Protection of the Ozone Layer discussed in Benedick, op. cit. note 3; industrial countries from Michael Grubb with Christiaan Vrolijk and Duncan Brack, *The Kyoto Protocol: A Guide and Assessment* (London: Royal Institute of International Affairs, 1999); quote from United Nations, op. cit. note 14.

16. Christopher Flavin, "Last Tango in Buenos Aires," *World Watch*, November/December 1998; "A Brief History of the Kyoto Protocol," in IISD, op. cit. note 5; Robert Engelman, *Profiles in Carbon: An Update on Population, Consumption and Carbon Dioxide Emissions* (Washington, DC: Population Action International, 1998).

17. "Kyoto Protocol to the United Nations Framework Convention on Climate Change," <www.unfccc.org/resource/docs/convkp/kpeng.html>, viewed 7 December 1999; on historic nature of Kyoto Protocol, see, for example, World Resources Institute, "WRI President Calls Kyoto Protocol a Historic Step for Humankind," press release (Washington, DC: 11 December 1997); problems with accord from Flavin, op. cit. note 16.

18. Emissions commitments from Grubb with Vrolijk and Brack, op. cit. note 15; emissions levels for 1990 and 1997 from U.N. Framework Convention on Climate Change (UNFCCC) Secretariat, "National Communications from Parties Included in Annex I to the Convention: Greenhouse Gas Inventory Data, 1990–1997" (Bonn: 29 September 1999); 1995 data from UNFCCC Secretariat, "National Communications from Parties Included in Annex I to the Convention: Annual Inventories of National Greenhouse Gas Data for 1996" (Bonn: 4 May 1999).

19. Sources for Table 6–1 are as follows: emissions commit-

ments from Grubb with Vrolijk and Brack, op. cit. note 15; emissions levels for 1990 and 1997 from UNFCCC Secretariat, "Inventory Data, 1990–1997," op cit. note 18; 1995 and 1996 data from UNFCCC Secretariat, "Inventories for 1996," op. cit. note 18. David G. Victor, Council on Foreign Relations, *Why The Kyoto Protocol and Greenhouse Gas Emissions Trading Will Fail, and How to Build a More Effective Global Warming Regime When They Do*, unpublished paper, 10 March 1999.

    20. Ibid.

    21. "U.S. Seeks Developing Country Participation Before Ratification of Climate Change Treaty," *International Environment Reporter*, 2 January 1998; agreement of 18 countries from United Nations, op. cit. note 6, viewed 7 December 1999; requirement of 55 countries from UNFCCC Secretariat, "Ministers Pledge to Finalize Climate Agreement by November 2000," press release (Bonn, Germany: 5 November 1999).

    22. Global Climate Information Project (GCIP), "Ad Campaign Addresses Concerns About Kyoto Protocol," press release, 8 September 1998, <www.sbsc.org/gw>, viewed 2 December 1999; members from GCIP, "GCIP Membership," <www.sbsc.org/gw/gcipmembership.htm>, viewed 2 December 1999; 13 million workers from <www.aflcio.org>, viewed 2 December 1999.

    23. Chinese and U.S. emissions from Christopher Flavin and Seth Dunn, *Rising Sun, Gathering Winds: Policies to Stabilize the Climate and Strengthen Economies*, Worldwatch Paper 138 (Washington DC: Worldwatch Institute, November 1997); quote from GGIP, "In Common," <www.sbsc.org/gw>, viewed 2 December 1999.

    24. Lee R. Raymond, Chairman and Chief Executive Officer, Exxon Corporation, "Energy—Key To Growth and a Better Environment for Asia-Pacific Nations," remarks at the World Petroleum Congress, Beijing, 13 October 1997.

    25. Benedick, op. cit. note 3; Rowlands, op.cit. note 3.

    26. Rowlands, op. cit. note 3; Benedick, op. cit. note 3.

    27. John J. Fialka, "Kyoto Treaty's Foes In U.S. Could Kill Pact Around the World," *Wall Street Journal*, 19 October 1999; Steve Liesman, "Big Business Produces Some Unexpected Converts," *Wall Street Journal*, 19 October 1999.

28. For an articulation of this concern, see Fialka, op.cit. note 27.

29. Benedick, op. cit. note 3; Armin Rosencranz and Reina Milligan, "CFC Abatement: The Needs of Developing Countries," *Ambio*, October 1990.

30. Benedick, op. cit. note 3; Richard J. Smith, chief U.S. negotiator for the London Amendments, "The Ozone Layer and Beyond—Towards a Global Environmental Diplomacy," remarks to the American Chemical Society, Washington, DC, 24 August 1994.

31. Benedick, op. cit. note 3; replenishments from Secretariat for the Multilateral Fund for the Implementation of the Montreal Protocol, "General Information," <www.unmfs.org>, viewed 2 December 1999.

32. Benedick, op. cit. note 3; Duncan Brack, *International Trade and the Montreal Protocol* (London: Earthscan, for Royal Institute of International Affairs, 1996).

33. UNEP, *1994 Report of the Economic Options Committee for the 1995 Assessment of the Montreal Protocol on Substances That Deplete the Ozone Layer* (Nairobi: 1994); current production and consumption from UNEP, op. cit. note 7.

34. COWIconsult, *Study on the Financial Mechanism of the Montreal Protocol* (Nairobi: UNEP, March 1995).

35. UNEP, op. cit. note 7; Multilateral Fund Secretariat, *Country Programme Summary Sheets* (Montreal: March 1999).

36. Nils Borg, International Association for Energy-Efficient Lighting, Berkeley, CA, e-mail to Molly O'Meara, Worldwatch Institute, 12 February 1998; Steven Nadel et al., *Lighting Energy Efficiency in China: Current Status, Future Directions* (Washington, DC: American Council for an Energy-Efficient Economy, May 1997).

37. Rakesh Bakshi, "Country Survey: India," *Wind Directions*, April 1997; "The Windicator," *Wind Power Monthly*, October 1999.

38. Examples of climate-friendly policies in developing countries from Center for Sustainable Development in the Americas and Worldwatch Institute, "Meaningful Action: A Proposal for Reducing Greenhouse Emissions & Spurring Energy Modernization in Developing Nations" (Washington, DC: unpublished paper, 21 October 1998); Christopher Flavin, "Banking Against Warming," *World Watch*, November/December 1997.

39. Projections for Chinese emissions compared with U.S. emissions from U.S. Department of Energy, *International Energy Outlook 1999* (Washington, DC: April 1999); recent emissions and economic growth trends in China from Christopher Flavin, "World Carbon Emissions Fall," *News Brief* (Washington, DC: Worldwatch Institute, 27 July 1999).

40. UNFCCC Secretariat, op. cit. note 21; Allen R. Myerson, "U.S. Splurging on Energy After Falling Off Its Diet," *New York Times*, 22 October 1998; sport utility vehicles from U.S. Public Interest Research Group, *Danger Ahead: Putting the Brakes on Global Warming Pollution* (Washington, DC: June 1999), and from David Kiley, "Not Your Father's SUV," *American Demographics*, January 1999; home sizes from Daniel Pedersen, "The Garage That Ate Tucson," *Newsweek*, 21 June 1999.

41. Benedick, op. cit. note 3.

42. Growth of wind and solar from Christopher Flavin, "Wind Power Blows to New Record," and Molly O'Meara, "Solar Cells Continue Double-Digit Growth," both in Brown, Renner, and Halweil, op. cit. note 10; International Energy Agency, *Indicators of Energy Use and Efficiency* (Paris: Organisation for Economic Co-operation and Development, 1997); Brad Knickerbocker, "Autos, 'Big Oil' Get Earth-Friendly," *Christian Science Monitor*, 6 October 1998.

43. Annual sales of $1 billion from John Browne, Group Chief Executive, The British Petroleum Company, "Climate Change, The Steps BP is Taking," presented at Stanford University, CA, 19 May 1997, <www.bpamoco.com/speeches/sp_970519. htm>, viewed 25 March 1999; $500 million from Jeroen van der Veer, Managing Director, Shell Group, and Jim Dawson, President, Shell International Renewables, "Shell International Renewables— Bringing Together the Group's Activities in Solarpower, Biomass and Forestry," presented 6 October 1997, <media.shell.com/ library/speech/0,1525,2303,00.html>, viewed 25 March 1999; 10 percent from Shell, "Taking Action on Sustainable Development: Climate Change," <www.shell.com/values/content/ 0,1240,1215-3397,00.html>, viewed 6 December 1999, and from BP Amoco, "Where BP Amoco Stands on Climate Change," <www.bpamoco.com/_nav/hse/index_climate.htm>, viewed 6

December 1999; John J. Fialka, "Global-Warming Treaty's Opposition Is Strained," *Wall Street Journal*, 30 October 1998; Mike R. Bowlin, "Clean Energy: Preparing Today for Tomorrow's Challenges," presented at Cambridge Energy Research Associates 18th Annual Executive Conference, *Globality & Energy: Strategies for the New Millennium*, Westin Galleria Hotel, Houston, TX, 9 February 1999.

44. John H. Cushman Jr., "Industrial Group Plans To Battle Climate Treaty," *New York Times*, 26 April 1998; Energy Innovations, *Energy Innovations: A Prosperous Path to a Clean Environment* (Washington, DC: Alliance to Save Energy, American Council for an Energy-Efficient Economy, Natural Resources Defense Council, Tellus Institute, and Union of Concerned Scientists, 1997); AFL-CIO Executive Council, "U.S. Energy Policy," policy statement (Miami, FL: 17 February 1999).

45. "1998 Global Surface Temperature Smashes Record," *Environment News Service*, <ens.lycos.com/ens/jan99/1999L-01-07-06.html>, 7 January 1999; Robert Marquand, "Glaciers in the Himalayas Melting at Rapid Rate," *Christian Science Monitor*, 5 November 1999; Curt Suplee, "Shrinkage Detected in Greenland's Ice," *Washington Post*, 5 March 1999; Joby Warrick, "Hot Year Was Killer for Coral," *Washington Post*, 5 March 1999; damage numbers from Munich Re, "Weather-Related Natural Disasters 1998" (Munich, Germany: 9 February 1999).

CHAPTER 7. TRADE WARS

1. "U.S. Imposes Sanctions in Beef Fight," *New York Times*, 19 July 1999; "EU Questions U.S.-Canadian Retaliation Request in Beef Dispute," *BRIDGES Weekly Trade News Digest*, 19 July 1999.

2. "Big Mac Targeted by French Farmers," *BRIDGES Weekly Trade News Digest*, 30 August 1999; Anne Swardson, "Something Is Rotten in Roquefort," *Washington Post*, 21 August 1999.

3. World Trade Organization (WTO), "Agreement on the Application of Sanitary and Phytosanitary Measures," in *Final Act: Agreement Establishing the World Trade Organization*, <www.wto.org/wto/legal/finalact.htm>, viewed 7 December 1999. See also Jeffrey S. Thomas and Michael A. Meyer, *The New Rules of Global*

Trade (Scarborough, ON, Canada: Carswell, 1997).

4. Thomas and Meyer, op. cit. note 3; Public Citizen, "Comments of Public Citizen, Inc. Regarding U.S. Preparations for the World Trade Organization's Ministerial Meeting Fourth Quarter 1999," comments to the U.S. Trade Representative, 22 October 1998, <www.consumerscouncil.org>, under "Other Policy Issues," viewed 7 December 1999.

5. James Cameron and Julie Abouchar, "The Status of the Precautionary Principle in International Law," in David Freestone and Ellen Hey, eds., *The Precautionary Principle and International Law: The Challenge of Implementation* (The Hague: Kluwer Law International, 1996); Rio Declaration from Lakshman Guruswamy, Geoffrey Palmer, and Burns Weston, *International Environmental Law and World Order* (St. Paul, MN: West Publishing, 1994); WTO provisions from WTO, op. cit. note 3, and from Thomas and Meyer, op. cit. note 3.

6. For the historical backdrop to the beef hormone case, see U.S. Office of Technology Assessment, *Trade and Environment: Conflicts and Opportunities* (Washington, DC: U.S. Government Printing Office, May 1992). For more recent events, see Paul Jacobs, "U.S., Europe Lock Horns in Beef Hormone Debate," *Los Angeles Times*, 9 April 1999, and Mark Suzman, "American Farmers Baffled as Europe Steers Clear of Beef Treated by Hormones," *Financial Times*, 22 July 1999. On possible harmful health effects of eating hormone-treated beef, see Michael Balter, "Scientific Cross-Claims Fly in Continuing Beef War," *Science*, 28 May 1999, and Dennis Bueckert, "Research Links Breast Cancer, Beef Hormones," *The Canadian Press*, 30 July 1999.

7. WTO, "EC Measures Concerning Meat and Meat Products (Hormones)," *Report of the Appellate Body* (Geneva: 16 January 1998); Public Citizen, op. cit. note 4.

8. Guy de Jonquières, "Genetically Modified Trade Wars," *Financial Times*, 18 February 1999; "EU Finalizes Labeling Rules for Genetically Modified Foods," *Business and the Environment*, July 1998; "Australia, NZ Require Mandatory GM Labels on Food," *Reuters*, 4 August 1999; "Japan Risks U.S. Ire With GMO Label Plan," *Reuters*, 5 August 1999; "South Korea to Start GMO Labelling from March 2001," *Reuters*, 30 November 1999; use of genetically modified organisms (GMOs) in U.S. food products from Rick Weiss, "In Europe, Cuisine du Gene Gets a Vehement Thumbs Down," *Washington Post*, 24 April 1999; on the move of European companies away from GMOs, see John Willman, "Consumer Power Forces Food Industry to Modify Approach," *Financial Times*, 10 June 1999, and Deborah Hargreaves, "Consumers' Unease Leads to Rethink on Modified Food Supplies," *Financial Times*, 29–30 May 1999; on the trade conflicts over labeling, see "Europe and US in Confrontation Over GM Food Labelling Criteria," *Nature*, 22 April 1999, and "U.S., Canada Concerned About Increase in Labeling Measures Affecting GMOs," *International Environment Reporter*, 23 June 1999.

9. Approval delays and drying up of European markets from Marian Burros, "U.S. Plans Long-Term Studies on Safety of Genetically Altered Foods," *New York Times*, 14 July 1999; on trends in Asia, see "Factbox—GMO Food Regulations in Asia," *Reuters*, 2 September 1999, and "EU Environment Ministers Virtually Ban Biotech Crops," *Environment News Service*, <ens.lycos.com/ens/jun99/1999L-06-25-02.html>, 25 June 1999.

10. For the U.S. government view, see Scott Kilman and Helene Cooper, "Crop Blight: Monsanto Falls Flat Trying to Sell Europe On Bioengineered Food," *Wall Street Journal*, 11 May 1999; "U.S. to Probe Euro Attitudes on GM Foods," *Environmental News Network*, <www.enn.com/news/wire-stories/1999/06/062599/eurogm_4022.asp>, 25 June 1999; and "TBT Committee Discusses Labelling, Standards," *BRIDGES Weekly Trade News Digest*, 14 June 1999. For a European perspective, see Friends of the Earth-UK, "GM Freeze: The Law's on Our Side," press release (London: 22 February 1999). On labeling, see "Sticky Labels," *The Economist*, 1 May 1999, and Matthew Stilwell and Brennan Van Dyke, *An Activist's Handbook on Genetically Modified Organisms and the WTO* (Washington, DC: The Consumer's Choice Council, March 1999).

11. On the breakdown of the biosafety talks, see Adam Thomson, "Efforts to Adopt UN Biosafety Protocol Fail," *Financial Times*, 25 February 1999, and "World Trade: Trade Tensions: The Biosafety Protocol Has Been Undermined by a Clash Between the Interests of US Multinationals and European Consumers," *Finan-*

cial Times, 26 February 1999; on continuing negotiations, see U.N. Environment Programme, "Negotiators to Seek Resumption of Biosafety Talks," press release (Nairobi: 15 September 1999).

12. "TBT Committee Discusses Labelling, Standards," op. cit. note 10; "U.S., Canada Call for GMO Trade on WTO Agenda," *BRIDGES Weekly Trade News Digest*, 10 May 1999; trade war warnings from de Jonquières, op. cit. note 8, and from David Wighton, "US Accused Over Modified Food Products," *Financial Times*, 29 September 1998.

13. General Agreement on Trade and Tariffs (GATT), *United States—Restrictions on Imports of Tuna: Report of the Panel* (Geneva: 3 September 1991).

14. On the history and results of the Marine Mammal Protection Act and its amendments, see Joshua R. Floum, Heller, Ehrman, White & McAuliffe, San Francisco, "GATT Challenge to Mexican Tuna Embargo," submission to Jane Early, U.S. National Oceanic and Atmospheric Administration (NOAA), 21 April 1991. See also Leesteffy Jenkins, "Using Trade Measures to Protect Biodiversity," in William J. Snape III, ed., *Biodiversity and the Law* (Washington, DC: Island Press, 1996).

15. Floum, op. cit. note 14; Jenkins, op. cit. note 14.

16. GATT, op. cit. note 13.

17. For a useful discussion of the clash between the trade and environment systems on the question of process standards, see the paper by Konrad Von Moltke in Sadruddin Aga Khan, ed., *Policing the Global Economy: Why, How, and For Whom?* Proceedings of an International Conference held in Geneva, March 1998 (London: Cameron May Ltd., 1998).

18. For a discussion of the tuna-dolphin ruling and its implications, see Eric Christensen and Samantha Geffin, "GATT Sets Its Net on Environmental Regulation: The GATT Panel Ruling on Mexican Yellowfin Tuna Imports and the Need for Reform of the International Trading System," *Inter-American Law Review*, winter 1991–1992; on the issue of conflicts between international environmental and trade agreements, see Steve Charnovitz, "Restraining the Use of Trade Measures in Multilateral Agreements: An Outline of the Issues," presentation at the Conference on the Relationship Between the Multilateral Trading System and the Use of

Trade Measures in Multilateral Agreements: Synergy or Friction? The Hague, 22–23 January 1996.

19. GATT, op. cit. note 13; Mexico's refusal to participate in 1991 agreement from Steve Charnovitz, "Dolphins and Tuna: An Analysis of the Second GATT Panel Report," *ELR News and Analysis* (Environmental Law Institute), October 1994.

20. Response of countries from David Phillips, "Dolphins and GATT," in Ralph Nader et al., *The Case Against Free Trade: GATT, NAFTA, and the Globalization of Corporate Power* (San Francisco and Berkeley: Earth Island Press and North Atlantic Books, 1993); dolphin mortality figures from NOAA, "Commerce Department Issues Initial Finding on Tuna/Dolphin Interactions; Will Adopt New Dolphin-Safe Label Standard," press release (Washington, DC: 29 April 1999).

21. On the role of unilateralism in paving the way for multilateral accords, see Steve Charnovitz, "Environmental Trade Measures: Multilateral or Unilateral?" *Environmental Policy and Law*, vol. 23, no. 3/4 (1993); on the fate of the tuna-dolphin panel report, see ibid. and Phillips, op. cit. note 20; on the EC challenge, see David Phillips and Mark J. Palmer, "Trade Agreement Runs Roughshod Over US Dolphin Protection Laws," in Peter Fugazzotto and Todd Steiner, *Slain by Trade: The Attack of the World Trade Organization on Sea Turtles and the US Endangered Species Act* (Forest Knolls, CA: Sea Turtle Restoration Project, July 1998); U.S. Department of State, "Multilateral Dolphin Protection Agreement Signed at Ceremony in Washington, D.C.," press release (Washington, DC: 21 May 1998).

22. "Agreement on the International Dolphin Conservation Program," 8 July 1998, <www.state.gov/www/global/oes/oceans/dolphin.html>, viewed 12 November 1999; requirement for dropping embargo from U.S. Department of State, op. cit. note 21.

23. "Meaning of Dolphin-Safe Tuna Label Changed," *Environment News Service*, <ens.lycos.com/ens/may99/1999L-05-04-04.html>, 4 May 1999.

24. Urgency and effectiveness of trade measures from Fugazzotto and Steiner, op. cit. note 21; 16 countries from David Hogan, Office of Marine Conservation, Bureau of Oceans and International Environmental and Scientific Affairs, U.S. Department of State, letter to author, 24 September 1999.

25. Adam Entous, "WTO Rules Against U.S. on Sea Turtle Protection Law," *Reuters*, 6 April 1998; Anne Swardson, "Turtle-Protection Law Overturned By WTO," *Washington Post*, 13 October 1998; WTO, *United States—Import Prohibition of Certain Shrimp and Shrimp Products* (Geneva: 12 October 1998). For a discussion of the new dispute resolution rules, see Thomas and Meyer, op. cit. note 3.

26. New guidelines from U.S. Department of State, "Revised Guidelines for the Implementation of Section 609 of Public Law 101-162 Relating to the Protection of Sea Turtles in Shrimp Trawl Fishing Operations," Public Notice 3086, *Federal Register*, 8 July 1999; prospects for a multilateral accord from Nancy Dunne, "Legal Wrangle Engulfs US Shrimp Dispute," *Financial Times*, 14 April 1999, and from Peter Fugazzatto, Sea Turtle Restoration Project, e-mail to Lisa Mastny, Worldwatch Institute, 29 July 1999; status of the shrimp-turtle case at the WTO from "Implementation Status of Shrimp-Turtle Ruling," *BRIDGES Between Trade and Sustainable Development*, July/August 1999.

27. "Trade Court Backs Protecting Turtles From Shrimpers' Nets," *Miami Herald*, 9 April 1999; Dunne, op. cit. note 26.

28. William Jefferson Clinton, "Remarks at the Commemoration of the 50th Anniversary of the World Trade Organization," Geneva, 18 May 1999; Group of Eight, "G-8 Communiqué Köln 1999," press release (Cologne, Germany: 20 June 1999).

29. For ideas on needed steps to reform the rules of world trade, see "The World Trade Organization and the Environment," Technical Statement by U.S. Environmental Organizations, 16 July 1999; National Wildlife Federation, *What's TRADE Got To Do With It?* (Washington, DC: 1999); World Wide Fund for Nature, *Sustainable Trade for a Living Planet: Reforming the World Trade Organization* (Gland, Switzerland: September 1999); and Lori Wallach and Michelle Sforza, *Whose Trade Organization?* (Washington, DC: Public Citizen, 1999). EU position from "EU Outlines Environmental Priorities for WTO's Next Round of Trade Talks," *International Environment Reporter*, 9 June 1999, and from "The EU Approach to the Millennium Round," Communication from the Commission to the Council and to the European Parliament,

<europa.eu.int/comm/trade/pdf/0807nr.pdf>, viewed 23 November 1999; U.S. position from "Trade and Sustainable Development," Communication from the United States, 30 July 1999, preparations for the 1999 Ministerial Conference, at <www.ustr. gov/new/gctandep.html>, and from "White House Declaration on Environmental Trade Policy," 16 November 1999, at <www.usia. gov/topical/econ/wto99/en1116d.htm>.

30. For a description of the agreement, see Thomas and Meyer, op. cit. note 3; for concerns about the agreement and prospects for the upcoming review, see Genetic Resources Action International, "Intellectual Property Rights and Biodiversity: The Economic Myths," *Global Trade and Biodiversity in Conflict*, October 1998, and David Downes and Matthew Stilwell, "The 1999 WTO Review of Life Patenting Under TRIPS," revised discussion paper (Washington, DC: Center for International Environmental Law, November 1998); for an update on the TRIPS discussions, see "North-South Divide Splits TRIPS Council," *BRIDGES Weekly Trade News Digest*, 25 October 1999.

31. Proposals to reduce subsidies and stance of EU and Japan from "Appeal to End Fishing Subsidies," *Financial Times*, 2 August 1999; $14–20 billion from Matteo Milazzo, *Subsidies in World Fisheries: A Reexamination*, World Bank Technical Paper No. 406, Fisheries Series (Washington, DC: World Bank, April 1998).

32. "Agriculture and the Environment: The Case of Export Subsidies," submission by Argentina et al. to the WTO's Committee on Trade and Environment (Geneva: 11 February 1999); stance of EU and Japan from Frances Williams, "World Trade: WTO Members Set to Sharpen Hatchets for Seattle Talks," *Financial Times*, 6 August 1999; energy subsidies from David Malin Roodman, *The Natural Wealth of Nations* (New York: W.W. Norton & Company, 1998); timber subsidies from Nigel Sizer, David Downes, and David Kaimowitz, "Tree Trade: Liberalization of International Commerce in Forest Products: Risks and Opportunities," *Forest Notes* (World Resources Institute and Center for International Environmental Law), November 1999.

33. On transparency issues, see Wallach and Sforza, op. cit. note 29; Daniel C. Esty, "Non-Governmental Organizations at the World Trade Organization: Cooperation, Competition, or Exclu-

sion," *Journal of International Economic Law*, March 1998; and
Gary P. Sampson, "Trade, Environment, and the WTO: A Frame-
work for Moving Forward," ODC Policy Paper (Washington, DC:
Overseas Development Council, February 1999).

34. "At Daggers Drawn," *The Economist*, 8 May 1999; paper
by Vandana Shiva in Aga Khan, op. cit. note 17.

35. John Burgess and Steven Pearlstein, "Protests Delay WTO
Opening," *Washington Post*, 1 December 1999; Suzanne Pardee,
"Demonstrators Swarm WTO Ministerial Meeting," *Environment
News Service*, <www.ens.lycos.com/ens/nov99/1999L-11-30-
01.html>, 30 November 1999.

CHAPTER 8. GREENING THE FINANCIAL ARCHITECTURE

1. On the reasons for the surge of private capital flows into
the developing world in the early 1990s, see World Bank, *Private
Capital Flows to Developing Countries* (New York: Oxford University Press, 1997), and Jacques de Larosière, "Financing Development in a World of Private Capital Flows: The Challenge for
Multilateral Development Banks in Working with the Private Sector," the Per Jacobsson Lecture, Washington, DC, September 1996.

2. Private finance totals and Figure 8–1 from World Bank,
Global Development Finance 1999 (Washington, DC: 1999), with
data deflated using the U.S. GNP Implicit Price Deflator provided
in U.S. Department of Commerce, *Survey of Current Business*, July
1999; $1.5 trillion from Bank for International Settlements, *Cen-
tral Bank Survey of Foreign Exchange and Derivatives Market Activity in April 1998* (Basel, Switzerland: May 1999).

3. History of crisis from World Bank, *East Asia: The Road to
Recovery* (Washington, DC: October 1998); capital outflows and
economic growth rates from International Monetary Fund (IMF),
World Economic Outlook, October 1999 (Washington, DC: 1999).

4. Russian default from Brian Duffy, "Market Chaos Goes
Global," *U.S. News and World Report*, 14 September 1998; Carlos
Lozado, "Brazilian Domino Effect?" *Christian Science Monitor*, 9
November 1998; $42-billion bailout from "The Real Thing," *The
Economist*, 21 November 1998.

5. Poverty and social fallout from World Bank, op. cit. note 3, from James D. Wolfensohn, President, World Bank Group, "The Other Crisis," Annual Meetings Address, Washington, DC, 6 October 1998, and from Kevin Sullivan, "A Generation's Future Goes Begging: Asia's Children Losing to Destitution," *Washington Post*, 7 September 1998; Peter Waldman, "Desperate Indonesians Devour Country's Trove of Endangered Species," *Wall Street Journal*, 26 October 1998; environmental budget cuts from World Bank, *Environment Matters, Annual Review* (Washington, DC: 1999), and from Stevie Emilia, "Crisis Forces Jakarta to 'Sacrifice' Its Environmental Programs," *Jakarta Post*, 2 July 1998.

6. Signs of economic recovery from IMF, op. cit. note 3; poverty rates from World Bank, *Poverty Trends and Voices of the Poor* (Washington, DC: 1999), and from Jean Michel Severino, Vice President, East Asia and Pacific Region, World Bank, "East Asia Regional Overview" (Washington, DC: 23 September 1999); future instability from Jeffrey E. Garten, "A Crisis Without a Reform," *New York Times*, 18 August 1999.

7. Jeffrey E. Garten, "Needed: A Fed for the World," *New York Times*, 23 September 1998; Dani Rodrik, "The Global Fix," *New Republic*, 2 November 1998.

8. See, for example, Jeffrey Sachs, "IMF Is a Power Unto Itself," *Financial Times*, 11 December 1997.

9. Total lending and adjustment lending from World Bank, *Annual Report 1999* (Washington, DC: 1999); policy reforms from Joseph Stiglitz, "More Instruments and Broader Goals: Moving Toward the Post-Washington Consensus," The 1988 WIDER Annual Lecture, Helsinki, Finland, 7 January 1998.

10. David Reed, ed., *Structural Adjustment, the Environment, and Sustainable Development* (London: Earthscan, 1996).

11. William D. Sunderlin, Consultative Group on International Agricultural Research, "Between Danger and Opportunity: Indonesia's Forests in an Era of Economic Crisis and Political Challenge" (Bogor, Indonesia: Center for International Forestry Research, 11 September 1998).

12. Asian cuts in environmental spending from World Bank, op. cit. note 5; Russia from Sara Zdeb, "Face the Facts! How the Global Economy Harms People and the Environment," informa-

tion sheet (Washington, DC: Friends of the Earth (FOE)–US, 16 April 1999); Brazil from Steven Schwartzman, Environmental Defense Fund, discussion with Lisa Mastny, Worldwatch Institute, 8 December 1999; high rates of deforestation from Daniel C. Nepstad et al., "Large-scale Impoverishment of Amazonian Forests by Logging and Fire," *Nature*, 8 April 1999.

13. "IMF Decides Once Again to Halt Disbursement of Loans Because of Illegal Timber Practices," *International Environment Reporter*, 15 October 1997; Government of Indonesia, "Supplementary Memorandum of Economic and Financial Policies: Fourth Review Under the Extended Agreement," document submitted to Michel Camdessus, IMF Managing Director, 16 March 1999, <www.imf.org/external/np/loi/1999/031699.htm>, viewed 19 October 1999; Sunderlin, op. cit. note 11.

14. Government of Indonesia, op. cit. note 13; Sunderlin, op. cit. note 11; abuses in reforestation fund from World Bank, *Environmental Implications of the Economic Crisis and Adjustment in East Asia* (Washington, DC: January 1999).

15. Andrea Durbin and Carol Welch, "Greening the Bretton Woods Institutions: Sustainable Development Recommendations for the World Bank and the International Monetary Fund" (Washington, DC: FOE–US, December 1998); Ved P. Gandhi, *The IMF and the Environment* (Washington, DC: IMF, 1998). I am indebted to Frances Seymour of the World Resources Institute for suggesting that the IMF's surveillance function might be expanded to cover environmental issues.

16. Evolving role of IMF from "Time For a Redesign?" *The Economist*, 30 January 1999, and from Carol Welch, "In Focus: The IMF and Good Governance," *U.S. Foreign Policy in Focus*, October 1988. See also Gandhi, op. cit. note 15.

17. World Bank policy on adjustment lending from "OD 8.60, Adjustment Lending Policy," in World Bank, *The World Bank Operational Manual*, <www.wbln0018.worldbank.org/Institutional/Manuals/OpManual.nsf>; World Bank, "The Evolution of Environmental Concerns in Adjustment Lending: A Review," paper prepared for the CIDIE Workshop on Environmental Impacts of Economywide Policies in Developing Countries, Washington, DC, 23–25 February 1993; less than 20 percent from

Nancy Dunne, "World Bank: Projects 'Fail' the Poor," *Financial Times*, 24 September 1999; World Bank environmental assessment policy as it applies to adjustment lending from Kathryn McPhail, World Bank, comment at meeting of NGOs with Ian Johnson, Vice President, Environmentally and Socially Sustainable Development, World Bank, 26 May 1999; failure of IMF to require environmental impact assessments from Durbin and Welch, op. cit. note 15.

18. Description of World Bank private-sector operations from World Bank, op. cit. note 9; 10 percent from World Bank, *Annual Report 1995* (Washington, DC: 1995). The World Bank has not updated this estimate in more recent annual reports.

19. International Finance Corporation (IFC), *Annual Report 1999* (Washington, DC: 1999).

20. For a description of the World Bank Group's environmental policies and guidelines, see World Bank, op. cit. note 12; IFC, "Procedure for Environmental and Social Review of Projects," information sheet (Washington, DC: December 1998); and Multilateral Investment Guarantee Agency, "Environmental and Social Review Procedures," <www.miga.org/disclose/soc_rev.htm>, viewed 4 November 1999. On importance of World Bank standards as a point of reference for private investors, see "Attorney Says Environmental Trends Will Have Far-Reaching Impact on U.S. Firms," *International Environment Reporter*, 1 November 1995, and World Bank, op. cit. note 5.

21. On the Bank's shortcomings in implementing policies, see Leyla Boulton, "World Bank Admits to Weakness on Environment," *Financial Times*, 4 October 1996, and Mark Suzman, "World Bank Accuses Itself of 'Serious Violations'," *Financial Times*, 7 January 1998; compliance unit and safeguards policies described in World Bank, op. cit. note 5.

22. U.S. Tibet Committee, "The World Bank: Undermining Tibet's Future, Executive Directors Approve China's Westward Colonization," e-mail to author, 16 August 1999; John Poirier, "World Bank to Probe China Loan," *Reuters*, 30 August 1999.

23. Environmental Defense Fund, *The Chad-Cameroon Oil and Pipeline Project: Putting People and the Environment at Risk* (Washington, DC: September 1999); "World Bank Completes Chad Oil Study, Waits for Consortium," *Reuters*, 12 November 1999.

24. Danielle Knight, "Finance-Environment: Increased Lending for Destructive Projects, *Inter Press Service*, 22 February 1999; Berne Declaration et al., *A Race to the Bottom* (Washington, DC: March 1999); export credit support in 1988 from Anthony Boote, *Official Financing for Developing Countries* (Washington, DC: IMF, 1995); support in 1996 from Anthony Boote and Doris Ross, *Official Financing for Developing Countries* (Washington, DC: IMF, 1998); 10 percent is a Berne Union estimate cited in Nancy Dunne, "Environmentalists Damn Export Credit Agencies' Policies," *Financial Times*, 31 July 1998.

25. Overseas Private Investment Corporation (OPIC), *OPIC Environmental Handbook* (Washington, DC: April 1999); U.S. Export-Import Bank, "Ex-Im Bank and the Environment," information sheet (Washington, DC: 8 June 1998).

26. Institute for Policy Studies, FOE–US, and International Trade Information Service, "OPIC, Ex-Im, and Climate Change: Business as Usual? An Analysis of U.S. Government Support for Fossil-Fueled Development Abroad, 1992–98" (Washington, DC: 28 April 1999); global emissions in 1998 based on data in Seth Dunn, "Carbon Emissions Dip," in Lester R. Brown, Michael Renner, and Brian Halweil, *Vital Signs 1999* (New York: W.W. Norton & Company, 1999).

27. Bolivia-Brazil pipeline from Lisa Mastny, "Death Sentence for a Tropical Forest," *World Watch*, November/December 1999, from Nancy Dunne, "OPIC Set to Approve Bolivian Pipeline Loan," *Financial Times*, 15 June 1999, and from OPIC, "OPIC Board Approves Cuiabá Project with Unprecedented Environmental Safeguards," press release (Washington, DC: 15 June 1999).

28. Three Gorges from Berne Declaration et al., op. cit. note 24; financing from "Breaking the Wall: China and the Three Gorges Dam," *Harvard International Review*, summer 1998; flooding and resettlement figures are U.S. Export-Import Bank estimates cited in "Ex-Im Bank Turns Down Requests to Consider Three Gorges Lending," *International Environment Reporter*, 12 June 1996; countries working on standards from Pam Foster, Halifax Initiative, Ottawa, ON, Canada, discussion with Lisa Mastny, Worldwatch Institute, 4 November 1999; international discussions on common standards from Group of Eight, "G-8 Communiqué Köln 1999," press release (Cologne, Germany: 20 June 1999).

29. Vanessa Houlder, "Greens Gun For Finance," *Financial Times*, 9 February 1999; Ann Monroe, "The Looming EcoWar," *Investment Dealers' Digest*, 24 May 1999; "Three Gorges Dam Sneaking Its Way Into Capital Markets," *The Bull and Bear Newsletter* (Quantum Leap Project, National Wildlife Federation and FOE–US), 20 January 1999.

30. Michelle Chan-Fishel, "Risk Exposure: Revealing Environmental and Political Risk to Private Financiers," draft report (Washington, DC: FOE–US, 11 May 1996); Bradford S. Gentry, *Private Investment and the Environment*, Discussion Paper 11 (New York: U.N. Development Programme, undated); Stephan Schmidheiny and Federico J.L. Zorraquín, with the World Business Council for Sustainable Development (WBCSD), *Financing Change* (Cambridge, MA: The MIT Press, 1996); "Bankers Urged to Include Assessments of Environmental Risk in Lending Decisions," *International Environment Reporter*, 14 May 1997.

31. Gentry, op. cit. note 30; John T. Ganzi and Julie Tanner, "Global Survey on Environmental Policies and Practices of the Financial Services Industry: The Private Sector," sponsored by National Wildlife Federation, Washington, DC, produced by Environment & Finance Enterprise, 16 May 1997.

32. U.N. Environment Programme (UNEP), "Bankers to Link Environment and Financial Performance," press release (New York: 16 May 1997); signatories and countries from "19 New Signatories to the Financial Services Initiatives," *The Bottom Line* (UNEP Financial Services Initiative newsletter), spring 1999, and from "New Signatories to the Initiatives," *The Bottom Line* (UNEP Financial Services Initiative newsletter), summer 1999; UNEP, "UNEP Statement by Financial Institutions on the Environment and Sustainable Development," revised version, May 1997, <www.unep.ch/eteu/finserv/english.html>, viewed 3 November 1999.

33. Abid Aslam, "Finance-Environment: Chinese Dam Tests 'Green' Banking Club," *Inter Press Service*, 13 September 1999; Julie Hill, Doreen Fedrigo, and Ingrid Marshall, *Banking on the Future, A Survey of Implementation of the UNEP Statement by Banks*

on Environment and Sustainable Development (London: Green Alliance, March 1997).

34. Claudia H. Deutsch, "For Wall Street, Increasing Evidence that Green Begets Green," *New York Times*, 19 July 1998; Edward Alden, "Go Green, Invest and Then Prosper," *Financial Times*, 25 January 1999.

35. Mark A. Cohen, Scott A. Fenn, and Jonathan S. Naimon, *Environmental and Financial Performance: Are They Related?* (Washington, DC: Investor Responsibility Research Center (IRRC), April 1995); Stanley J. Feldman, Peter A. Soyka, and Paul Ameer, *Does Improving a Firm's Environmental Management System and Environmental Performance Result in a Higher Stock Price?* (Fairfax, VA: ICF Kaiser International, Inc., November 1996); see also Jerald Blumburg, Åge Korsvold, and Georges Blum, *Environmental Performance and Shareholder Value* (Geneva: WBCSD, undated), and "Environmentally Responsible Investments Outperform Others," *Business and the Environment*, July 1997.

36. Global Environment Fund, "The Global Environment Fund Group, 1998/9," information sheet (Washington, DC: October 1998); 10 countries from Greg Nagler, Investment Analyst, Global Environment Fund, discussion with Payal Sampat, Worldwatch Institute, 30 October 1997; John Ganzi, Sandy Buffett, and Robin Dunn, "A Review of Publicly Available Funds That Focus on Financial and Environmental Performance," report prepared by Environment & Finance Enterprise, for the Office of Cooperative Environmental Management, U.S. Environmental Protection Agency, November 1998.

37. Social Investment Forum, *1999 Report on Socially Responsible Investing Trends in the United States* (Washington, DC: 4 November 1999); Domini 400 Index from Joanne Legomsky, "Investing By Conscience is Paying Off These Days," *New York Times*, 24 January 1999, and from KLD, "Domini 400 Social Index Gains 6.86% in October, S&P 500 Gains 6.36%," preliminary press release (Cambridge, MA: 9 November 1999).

38. Dow Jones Sustainability Group Index, "Dow Jones Indexes and SAM Sustainability Group Launch Sustainability Indexes," press release (Zurich: 8 September 1999); "Europe Tops Business Sustainability Index," *Environment News Service*,

<www.ens.lycos.com/ens/sep99/1999L-09-20-01.html>, 20 September 1999.

39. Social Investment Forum, op. cit. note 37.

40. "Global Reporting Initiative," background information available at <www.ceres.org/reporting/globalreporting.html>; "Ceres Green Reporting Guidelines Launched," *Environment News Service*, 3 March 1999; for more information, see <www.global reporting.org>.

41. Schmidheiny and Zorraquín, op. cit. note 30; IRRC, "Environmental Disclosure at S&P 500 Companies' Non-U.S. Operations," press release (Washington, DC: 1 December 1997); Michelle Chan-Fishel, "The Corporate Sunshine Working Group: Expanding SEC Disclosure Requirements for Corporate Accountability" (Washington, DC: FOE–US, undated); Michelle Chan-Fishel, FOE–US, Washington, DC, e-mail to author, 1 December 1999.

42. Patareeya Benjapolchai, Senior Vice President, Stock Exchange of Thailand, "Stock Exchange Policies for Protecting the Environment," address to The Environment and Financial Performance, UNEP's Third International Roundtable Meeting on Finance and the Environment, Columbia University, New York, 22–23 May 1997.

43. "Strengthening the International Financial Architecture," Report of G7 Finance Ministers to the Köln Economic Summit, Cologne, Germany, 18–20 June 1999. For ideas on integrating environmental issues into the financial system, see John Ganzi, Frances Seymour, and Sandy Buffett, *Leverage for the Environment* (Washington, DC: World Resources Institute, 1998).

CHAPTER 9. STRENGTHENING GLOBAL ENVIRONMENTAL GOVERNANCE

1. Treaties and Figure 9–1 are compiled from U.N. Environment Programme (UNEP), *Register of International Treaties and Other Agreements in the Field of the Environment 1996* (Nairobi: 1996), from United Nations, *The United Nations Treaty Collection*, electronic database, <www.un.org/Depts/Treaty>, viewed 9 March 1999, and from UNEP, "International Conventions and Protocols in the Field of the Environment: Report of the Executive Director," prepared for UNEP Governing Council Twentieth Session, 13 November 1998.

2. Table 9–1 is based on the following sources: number of parties for Law of the Sea, Montreal Protocol, fish stocks agreement, and the Basel, climate change, biological diversity, desertification, and prior informed consent (PIC) conventions from United Nations, op. cit. note 1, viewed 2 December 1999; International Whaling Commission (IWC), "Membership," <ourworld. compuserve.com/homepages/iwcoffice/IWC.htm#Members>, viewed 2 December 1999; whaling convention from IWC, "History and Purpose," and "Conservation and Management," information sheets, <ourworld.compuserve.com/homepages/iwcoffice>, viewed 8 July 1999; Antarctic treaty and parties from The Antarctica Project, "The Antarctic Treaty System," <www.asoc.org/general/ats.htm>, viewed 2 December 1999; Convention on International Trade in Endangered Species of Wild Fauna and Flora (CITES) Secretariat, "What is CITES?," <www.cites.org/CITES/eng/what-is.shtml>, viewed 2 December 1999; CITES and ivory ban from Caroline Taylor, "The Challenge of African Elephant Conservation," *Conservation Issues* (World Wildlife Fund (WWF-US)), April 1997; Law of the Sea from IUCN–World Conservation Union, "The Law of the Sea: Priorities and Responsibilities in Implementing the Convention" (Gland, Switzerland: 1995); Montreal Protocol from UNEP, "Press Backgrounder: Basic Facts and Data on the Science and Politics of Ozone Protection," information sheet (Nairobi: May 1999); Basel Convention from UNEP, "Public Information Leaflets on the Basel Convention," <www.unep.ch/basel/index.html>, viewed 8 July 1999; Basel amendment from UNEP, "Basel Convention Takes Stronger Stand on Waste," *UNEP Update*, November 1995; climate change convention from UNEP, "Climate Change Information Sheet 18: The Climate Change Convention" and "Climate Change Information Sheet 21: The Kyoto Protocol," <www.unep.ch/iuc>, viewed 8 July 1999; biodiversity convention from UNEP, "Convention on Biological Diversity," <www.unep.ch/iuc>, viewed 8 July 1999; desertification convention from Secretariat of the Convention to Combat Desertification, "Fact Sheet 1—An Introduction to the United Nations Convention

to Combat Desertification," <www.unccd.ch>, viewed 8 July 1999; fish stocks agreement from Giselle Vigneron, "The Most Recent Efforts in the International Community to Implement the 1995 United Nations Straddling Fish Stocks Agreement," *Colorado Journal of International Environmental Law and Policy*, Yearbook 1998; prior informed consent convention from Cheryl Hogue, "Treaty on Prior Informed Consent Signed by 57 Countries at Rotterdam Meeting," *International Environment Reporter*, 16 September 1999.

3. Gareth Porter and Janet Welsh Brown, *Global Environmental Politics* (Boulder, CO: Westview Press, 1996); air pollution treaty from Marc Levy, "European Acid Rain: The Power of Tote-Board Diplomacy," *Institutions for the Earth: Sources of Effective International Environmental Protection* (Cambridge, MA: The MIT Press, 1993); chlorofluorocarbon (CFC) production in 1988 from Sharon Getamel, DuPont, Wilmington, DE, letter to Worldwatch, 15 February 1996; CFC production in 1997 from UNEP, *Data Report on Production and Consumption of Ozone Depleting Substances, 1986–1998* (Nairobi: 1999); CITES from Ronald Orenstein, "Africa's Elephants Could Soon Be Under the Gun Again," *Christian Science Monitor*, 2 February 1992; whale takes in 1961 from Elizabeth Kemf and Cassandra Phillips, *Wanted Alive! Whales in the Wild* (Gland, Switzerland: World Wide Fund for Nature (WWF-International), October 1995); current whale takes a rough estimate calculated from 415 takes allotted to native whalers, from IWC, "Catch Limits For Aboriginal Subsistence Whaling," <ourworld.compuserve.com/homepages/iwcoffice/ Catches.htm#Commercial>, viewed 2 December 1999, and from 1,100 additional takes, compiled from WWF-International, "Japanese Fleet Returns from Whale Sanctuary with 389 Minkes," press release (Gland, Switzerland: 23 April 1999), and from Kieran Mulvaney, "The Whaling Effect," news feature (Gland, Switzerland: WWF-International, October 1999); The Antarctica Project, "The Protocol on Environmental Protection to the Antarctic Treaty," 14 June 1999, <www.asoc.org/currentpress/protocol.htm>, viewed 3 November 1999.

4. On the role of transparency, see Abram Chayes and Antonia Handler Chayes, *The New Sovereignty* (Cambridge, MA: Harvard University Press, 1995); Ronald B. Mitchell, "Sources of

Transparency: Information Systems in International Regimes,"
International Studies Quarterly, vol. 42 (1998); and U.S. General
Accounting Office (GAO), *International Environment: Literature on
the Effectiveness of International Environmental Agreements* (Washington, DC: May 1999).

5. GAO, op. cit. note 4; biodiversity convention from UNEP,
"Synthesis of Information Contained in National Reports on the
Implementation of the Convention," 30 April 1998, Item 11 of the
Provisional Agenda of the Fourth Meeting of the Conference of the
Parties to the Convention on Biological Diversity, Bratislava, 4–15
May 1998; U.N. Framework Convention on Climate Change,
"Review of the Implementation of Commitments and of Other
Provisions of the Convention: National Communications from
Parties Included in Annex I to the Convention," 18 September
1998, Item 4 of the Provisional Agenda of the Fourth Session of
the Conference of the Parties, Buenos Aires, 2–13 November 1998;
UNEP, "Report of the Secretariat on Information Provided by the
Parties in Accordance with Article 7 and 9 of the Montreal Protocol: The Reporting of Data by the Parties to the Montreal Protocol
on Substances That Deplete the Ozone Layer," 21 September 1998,
prepared for the Tenth Meeting of the Conference of the Parties to
the Montreal Protocol on Substances That Deplete the Ozone
Layer, Cairo, 23–24 November 1998.

6. UNEP, *Global Environmental Outlook 2000* (Nairobi: 1999).

7. GAO, op. cit. note 4; Rosemary Sandford, "International
Environmental Treaty Secretariats: a Case of Neglected Potential?"
Environmental Impact Assessment Review, vol. 198, no. 1 (1996);
the 2000 budget for the Ramsar convention is only about $1.9 million, per "Financial and Budgetary Matters," Resolution No. VII.28
of 7th Meeting of the Conference of the Contracting Parties, Convention on Wetlands (Ramsar), San José, Costa Rica, 10–18 May
1999, while the 2000 budget for the climate convention is $11 million, per U.N. Framework Convention on Climate Change,
"Administrative and Financial Matters: Programme Budget for the
Biennium 2000–2001," Item 8 of the provisional agenda of the
Fifth Session of the Conference of the Parties, Bonn, Germany, 25
October–5 November 1999; the Clean and Safe Water program of
the U.S. Environmental Protection Agency (EPA), for instance,

had a budget of $942 million in 1999, per EPA, "Summary of the 2000 Budget" (Washington, DC: January 1999).

8. Sandford, op. cit. note 7; Gordon Binder and Jonathan Adams, "Does CITES Need More Teeth," *Conservation Issues* (WWF-US), October 1994.

9. Ozone treaty Secretariat from <www.unep.org/ozone/about.htm>, viewed 8 December 1999; climate treaty Secretariat from <www.unfccc.org/secret/index.html>, viewed 8 December 1999; biodiversity convention secretariat from <www.biodiv.org/SEC.HTML>, viewed 8 December 1999; United Nations, *Agenda 21: The United Nations Program of Action from Rio* (New York: 1992); Rosemary Sandford, "International Environmental Treaty Secretariats: Stage-Hands or Actors?" in Helge Ole Bergesen and Georg Parmann, eds., *Green Globe Yearbook 1994* (Oxford: Oxford University Press, for the Fridtjof Nansen Institute, 1994).

10. For a general description of Pelly Amendment as well as sea turtle example, see Steve Charnovitz, "Encouraging Environmental Cooperation Through the Pelly Amendment," *Journal of Environment and Development*, winter 1994 (CITES included the sea turtles at issue on Appendix 1, but Japan had taken a reservation, so it was not technically in legal violation of the accord); "U.S. Bans Importing Fish Caught With Driftnets," *Washington Post*, 20 September 1991; Tom Kenworthy, "Japan to End Drift Net Fishing in Bow to Worldwide Pressure," *Washington Post*, 27 November 1991.

11. Discussion of Montreal, Basel, and CITES trade provisions in UNEP, *Policy Effectiveness and Multilateral Environmental Agreements*, Environment and Trade Series No. 17 (Geneva: Economics, Trade, and Environment Unit, 1998); on the question of whether trade provisions induced widespread participation in Montreal, see Duncan Brack, *International Trade and the Montreal Protocol* (London: Earthscan, for the Royal Institute of International Affairs, 1996); CITES recommendations on China, Italy, and Thailand from Peter H. Sand, "Commodity or Taboo? International Regulation of Trade in Endangered Species," in Helge Ole Bergesen and Georg Parmann, eds., *Green Globe Yearbook 1997* (Oxford: Oxford University Press, for the Fridtjof Nansen Institute, 1997); recommendations on Taiwan from "Taipei to Meet CITES Requirements," *Taipei CNA*, 21 April 1994; recommendations on Greece from TRAFFIC International, "Wildlife Trade Ban Called for Against Greece," press release (Cambridge, U.K.: 20 August 1998); U.S. ban from "Government Will Not Oppose U.S. Ban on Thai Wildlife Imports," *The Nation* (Bangkok), 5 July 1991; Thomas L. Friedman, "U.S. Puts Sanctions on Taiwan," *New York Times*, 12 April 1994.

12. For descriptions of the discussions in the Committee on Trade and Environment, see World Trade Organization (WTO), "Background Document for High-Level Symposium on Trade and Environment" (Geneva: 15–16 March 1999), and Gary P. Sampson, "Trade, Environment, and the WTO: A Framework for Moving Forward" (Washington, DC: Overseas Development Council, February 1999).

13. For a proposal to insulate multilateral environmental agreements from challenge at the WTO, see WWF-International, Greenpeace, and Friends of the Earth, "Joint Statement on the Relationship Between the WTO and Multilateral Environmental Agreements," position statement (Gland (Switzerland), Amsterdam, and London: July 1996); Article 104 in Governments of Canada, Mexico, and the United States of America, "The North American Free Trade Agreement," 6 September 1992; "EC Blocks Trade Threats to Environmental Agreements," *ENDS Report*, November 1996; on "savings clauses," otherwise known as "supremacy clauses," see Center for International Environmental Law, "WTO 'Supremacy Clause' in the POPs Convention," e-mail to author, 7 July 1999.

14. Global Environment Facility (GEF), brochure (Washington, DC: undated); "Agreement Reached on Funding GEF; Program to Receive More than $2 Billion," *International Environment Reporter*, 23 March 1994; 1998 replenishment from GEF, "Introduction to the GEF," informational brochure, <www.gefweb.org/intro/gefintro.pdf>, viewed 22 November 1999; allocations as of 1999 from GEF, "Draft Annual Report 1999: Volume 1" (Washington, DC: 22 November 1999). Countries eligible for ozone project funding under the Interim Multilateral Fund are ineligible for GEF support for this purpose. The main beneficiaries of GEF support for ozone projects are countries from the former Soviet Union and

Eastern Europe.

15. Roles of implementing agencies from GEF, brochure, op. cit. note 14; collaboration with others in executing projects from GEF Council, "Expanded Opportunities for Executing Agencies" (Washington, DC: 10 September 1998).

16. GEF, "Instrument for the Establishment of the Restructured Global Environment Facility," Report of the GEF Participants Meeting, Geneva, Switzerland, 14–16 March 1994.

17. "Summary Report of the First Assembly of the Global Environment Facility," *Sustainable Developments*, 4 April 1998; *Study of GEF's Overall Performance*, document presented to the GEF Assembly, New Delhi, 2 March 1998.

18. *Study of GEF's Overall Performance*, op. cit. note 17.

19. Ibid.; spending on fossil fuel projects versus GEF budget from "GEF Deemed Success at First Assembly; Funding to be Increased by 30 Percent," *International Environment Reporter*, 15 April 1998.

20. "GEF to Work Closely With Private Sector In Next Four Years, Fund Officials Say," *International Environment Reporter*, 15 April 1998; Dana Younger, International Finance Corporation (IFC), e-mail to author, 15 December 1999; GEF, "Private Sector Information Kiosk," <www.gefweb.org/PRIVATE/priv.htm>, viewed 8 December 1999; Douglas Salloum, Small and Medium Enterprises Program Manager, IFC, discussion with Lisa Mastny, Worldwatch Institute, 9 December 1999.

21. On both the promise and the potential pitfalls of the Clean Development Mechanism, see Seth Dunn, "Can North and South Get in Step?" *World Watch*, November/December 1998, and Michael Grubb with Christiaan Vrolijk and Duncan Brack, *The Kyoto Protocol: A Guide and Assessment* (London: Royal Institute of International Affairs, 1999).

22. Konrad von Moltke, "Why UNEP Matters," A Report for WWF-International (Gland, Switzerland: March 1995); UNEP, "UNEP: Two Decades of Achievement and Challenge" (Nairobi: October 1992).

23. UNEP spending over its first 20 years from von Moltke, op. cit. note 22; the World Wildlife Fund's total expenses in 1999 were nearly $97 million, per WWF-US, *Annual Report 1999*

(Washington, DC: 1999). Table 9–2 derived from the following sources: International Monetary Fund (IMF) staff and total disbursements from IMF, *Annual Report 1999* (Washington, DC: 1999); World Bank staff and lending commitments from World Bank, *Annual Report 1999* (Washington, DC: 1999); U.N. Development Programme (UNDP) staff from UNDP, *UNDP Today: Introducing the Organization* (New York: September 1998); UNDP total expenditures from UNDP, "Annual Report of the Administrator for 1998 and Related Matters: Statistical Annex," Item 2 of the Provisional Agenda for the Annual Session of the Executive Board, New York, 14 May 1999; UNICEF staff and total expenditure from UNICEF, *1999 UNICEF Annual Report* (New York: 1999); World Health Organization (WHO) staff from WHO, "Proposed Programme Budget for 2000–2001," Item 12 of the Provisional Agenda for the Fifty-Second World Health Assembly, Geneva, 21 April 1999; WHO total integrated budget from WHO, "What is the Budget of the WHO?" <www.who.int/aboutwho/en/qa6.htm>, viewed 24 June 1999; U.N. Food and Agriculture Organization (FAO) staff and total estimated expenditure from FAO, "Summary Programme of Work and Budget 2000–01," prepared for the 116th Session of the FAO Council, Rome, 14–19 June 1999; UNESCO staff from UNESCO, "How It Works," information sheet, <www.unesco.org/general/eng/about/how.html>, viewed 24 June 1999; UNESCO resources (regular budget plus extrabudgetary resources) from UNESCO, "Approved Programme and Budget for 1998 and 1999," Document 29 C/5 (Paris: 1997); U.N. Population Fund (UNFPA) staff and total expenditures from UNFPA, "Report of the Executive Director for 1998: Statistical Overview," Item 11 of the Provisional Agenda for the 1999 Annual Session of the Executive Board, New York, 14–23 June 1999; International Labour Organisation (ILO) staff from ILO, Programme, Financial and Administrative Committee (PFAC), "Composition and Structure of the Staff," Item 11 on the Agenda for the 274th Session of the Governing Body, Geneva, March 1999; ILO expenditure from ILO, PFAC, "Programme and Budget for 1998–99: Position of Accounts as at 31 December 1998," Item 1 on the Agenda for the 274th Session of the Governing Body, Geneva, March 1999; International Atomic Energy Agency (IAEA) resources (regular budget plus extrabudgetary contributions) from IAEA, *Annual Report 1998* (New York: 1998); UNEP total resources from UNEP, "Programme, the Environment Fund and Administrative and Other Budgetary Matters," report of the Executive Director to the 20th Session of the UNEP Governing Council, Nairobi, 1–5 February 1999; WTO total budget from WTO, *Annual Report 1998* (Geneva: 1998); U.N. Industrial Development Organization (UNIDO) staff from "UNIDO at a Glance," information sheet, <www.unido.org/doc/66.html>, viewed 7 December 1999; UNIDO expenditure (disbursements plus technical cooperation delivery) from UNIDO, "Financial Situation of UNIDO," Item 3 of the provisional agenda for the 15th Session of the Programme and Budget Committee, Vienna, 14–16 April 1999, prepared 13 February 1999; World Meteorological Organization (WMO) staff and resources (regular budget plus extrabudgetary resources) from "Basic Facts About WMO," information sheet, <www.wmo.ch/web-en/wmofact.html>, viewed 24 June 1999; International Maritime Organization total budget from "Frequently Asked Questions About IMO," information sheet, <www.imo.org/imo/faqs.htm>, viewed 24 June 1999.

24. UNEP, *Twenty Years Since Stockholm*, 1992 Annual Report of the Executive Director (Nairobi: 1993); UNEP, *Annual Report 1998* (Nairobi: January 1999); regional seas from Halifa O. Drammeh, UNEP, e-mail to Lisa Mastny, Worldwatch Institute, 30 October 1998; UNEP Division of Technology, Industry, and Economics from <www.unepie.org>, viewed 8 December 1999; for a list of UNEP regional offices, see <www.unep.org/earthw/unepwww.htm>.

25. "Nairobi Declaration on the Role and Mandate of the United Nations Environmental Programme, Adopted 7 February 1997, UNEP, Nairobi," reprinted in UNEP, *Annual Report*, op. cit. note 24; UNEP Regional Office for North America, "UN General Assembly Elects Klaus Topfer New Executive Director of UN Environment Programme," press release (New York: 1 February 1998); update on UNEP's restructuring from Earth Negotiations Bulletin, "20th Session of the Governing Council of the United Nations Environment Programme: 1–5 February 1999" (Winnipeg, MN, Canada: International Institute for Sustainable Development, 8

February 1999), and from UNEP, *Annual Report*, op. cit. note 24.

26. For a prominent proposal to create a global environmental organization, see Daniel C. Esty, "GATTing the Greens," *Foreign Affairs*, November/December 1993, and Daniel C. Esty, "The Case for a Global Environmental Organization," in Peter B. Kenen, ed., *Managing the World Economy: Fifty Years After Bretton Woods* (Washington, DC: Institute for International Economics, 1994). See also Frank Biermann and Udo E. Simonis, "A World Environment and Development Organization," Policy Paper 9 (Bonn: Development and Peace Foundation/Stiftung Entwicklung und Frieden, June 1998). For a discussion of the debate in the 1970s about both the merits and disadvantages of an independent international environment agency, see George F. Kennan, "To Prevent a World Wasteland," *Foreign Affairs*, April 1970. Kennan himself strongly advocated the value of an independent agency at that time. Both James Gustave Speth, former Administrator of UNDP, and Renato Ruggiero, former Director-General of the WTO, have voiced support for creating a World Environment Organization over the last few years. See James Gustave Speth, "Note Regarding Questions on UN Reform Proposals," unpublished document, 4 October 1996, and James Gustave Speth, Administrator, UNDP, "Address at the United Nations General Assembly Special Session for the Overall Review and Appraisal of the Implementation of Agenda 21," New York, 26 June 1997; Renato Ruggiero, "Opening Remarks to the High Level Symposium on Trade and the Environment," 15 March 1999, <www.wto.org/wto/hlms/dgenv.htm>, viewed 14 April 1999.

27. "French President Calls for Establishment of Global Unit to Link Environmental Treaties," *International Environment Reporter*, 11 November 1998; "Strengthening the Role of the United Nations Environment Programme in Promoting Collaboration among Environmental Conventions," Report of the Executive Director, Twentieth Session of the UNEP Governing Council, Nairobi, 7 January 1999; "Summary Report of Inter-Linkages— International Conference on Synergies and Coordination Between Multilateral Environmental Agreements, 14–16 July 1999," *Sustainable Developments*, 18 July 1999.

28. James Avery Joyce, *World Labor Rights and Their Protec-*

tion (London: Croom Helm, 1980); Steve Charnovitz, "Promoting Higher Labor Standards," *The Washington Quarterly*, summer 1995; Steve Charnovitz, "Improving Environmental and Trade Governance," *International Environmental Affairs*, winter 1995.

29. United Nations, op. cit. note 9; UNDP, "A Guide to UNDP's Sustainable Energy & Environment Division," <www.undp.org/seed/guide/intro.htm>, viewed 8 December 1999.

30. For a list of U.N. specialized agencies, see "Official Web Site Locator for the United Nations System of Organizations," <www.unsystem.org/index2.html>, viewed 8 December 1999; WMO from <www.wmo.ch>, viewed 8 December 1999; WHO from <www.who.ch>, viewed 8 December 1999; on WHO's standard-setting role, see, for instance, WHO, *Air Quality Guidelines for Europe* (Copenhagen: 1987); FAO from <www.fao.org>, viewed 8 December 1999; UNFPA and Cairo Action Plan from UNFPA, "Coming Up Short: Struggling to Implement the Cairo Programme of Action," <www.unfpa.org/modules/intercenter/upshort/index.htm>, viewed 8 December 1999.

31. Kathryn G. Sessions, "Institutionalizing the Earth Summit," UNA-USA Occasional Paper (Washington, DC: United Nations Association of the United States of America, October 1992).

32. Earth Council, *National Councils for Sustainable Development Database*, electronic database, <www.ecouncil.ac.cr/database>, viewed 9 December 1999; Sheila Henry, International Council for Local Environmental Initiatives, Toronto, ON, Canada, discussion with Lisa Mastny, Worldwatch Institute, 9 December 1999.

33. Tom Biggs and Felix Dodds, "The UN Commission on Sustainable Development," in Felix Dodds, ed., *The Way Forward* (London: Earthscan, 1997); "Greater Private-Sector Role Suggested in U.N. Work on Production, Consumption," *International Environment Reporter*, 3 March 1999; indicators from United Nations, Department for Policy Coordination and Sustainable Development, Division for Sustainable Development, "Indicators of Sustainable Development: Framework and Methodologies," <www.un.org/esa/sustdev/newspubs.htm>, viewed 9 December 1999; review of industry initiatives from International Institute for

Sustainable Development, "The Sixth Session of the Commission on Sustainable Development, 20 April–1 May 1998," *Earth Negotiations Bulletin*, 4 May 1998.

34. Number of nongovernmental representatives from Zehra Aydin-Sipos, Major Groups Focal Point, U.N. Commission on Sustainable Development, discussion with Lisa Mastny, Worldwatch Institute, 9 December 1999; "Culture Change in Tone Noted At Seventh Session of U.N. CSD," *International Environment Reporter*, 12 May 1999.

CHAPTER 10. PARTNERSHIPS FOR THE PLANET

1. Mike Dolan quoted in Greg Miller, "WTO Summit: Protest in Seattle; Internet Fueled Global Interest in Disruptions," *Los Angeles Times*, 2 December 1999.

2. Joseph Kahn and David E. Sanger, "Trade Obstacles Unmoved, Seattle Talks End in Failure," *New York Times*, 4 December 1999; for the distinction between governance and governments, see Commission on Global Governance, *Our Global Neighborhood* (New York: Oxford University Press, 1995), and Wolfgang H. Reinicke, *Global Public Policy* (Washington, DC: Brookings Institutions Press, 1998).

3. Thomas Princen and Matthias Finger, *Environmental NGOs in World Politics* (London: Routledge, 1994); Curtis Runyan, "The Third Force: NGOs," *World Watch*, November/December 1999; Helmut K. Anheier and Lester M. Salamon, eds., *The Nonprofit Sector in the Developing World* (New York: Manchester University Press, 1998).

4. Historical information and Figure 10–1 based on Union of International Associations, *Yearbook of International Organizations* (Munich: K.G. Saur Verlag, various years); environmental nongovernmental organizations (NGOs) as share of total from Margaret E. Keck and Kathryn Sikkink, *Activists Beyond Borders* (Ithaca, NY: Cornell University Press, 1998), based on data from the Union of International Associations.

5. World Wide Fund for Nature (WWF International), "History," <www.panda.org/wwf/history/history.htm>, viewed 9

December 1999; Greenpeace International, "Join," <www. greenpeace.org/join.shtml>, viewed 9 December 1999. Table 10–1 derived from the following sources: BirdLife International, <wnnserv.wnn.or.jp/wnn-n/w-bird/bli-c.html>, viewed 14 December 1999; International Federation of Organic Agriculture Movements (IFOAM) from J. Patrick Madden and Scott G. Chaplowe, *For ALL Generations: Making World Agriculture More Sustainable* (Glendale, CA: OM Publishing, 1997); Pesticide Action Network North America, "What is PANNA?" <www.panna.org/panna/ whatis.html>, viewed 1 July 1999; International Rivers Network, "About IRN," <www.irn.org>, viewed 1 July 1999; Climate Action Network, "Climate Action Network: A Force for Change," <www.climatenetwork.org>, viewed 6 July 1999; Women's Environment & Development Organization, "About WEDO," <www.wedo.org/about/about.htm>, viewed 1 July 1999; Biodiversity Action Network, "Who We Are and What We Do," <www.bionet-us.org/who-we-r.html>, viewed 6 July 1999; International NGO Network on Desertification and Drought, "RIOD—An Introduction," <riod.utando.com>, viewed 1 July 1999; World Forum of Fish Workers & Fish Harvesters," <www.south-asian-initiative.org/wff/intro.htm>, viewed 25 June 1999; International POPs Elimination Network, "Organizational Structure," <www.psr.org/ipen/ipen_structure.htm>, viewed 1 July 1999.

6. For the role of activists in overturning the Multilateral Agreement on Investment, see Madelaine Drohan, "How the Net Killed the MAI," *Toronto Globe and Mail*, 29 April 1998, and R.C. Longworth, "Activists on Internet Reshaping Rules of Global Economy," *Chicago Tribune*, 5 July 1999; for the environmental flaws in the agreement, see WWF International, "The OECD Multilateral Agreement on Investment," WWF International Briefing (Gland, Switzerland: March 1997), and Friends of the Earth–US, "The OECD Multilateral Agreement on Investment (MAI): Examples of Laws that Would Conflict with the MAI," <www.foe.org/ga/exshort.html>, viewed 23 October 1997.

7. *Brent Spar* incident described in Control Risks Group, *No Hiding Place: Business and the Politics of Pressure* (London: July 1997); MacMillan Bloedel from Edward Alden, "MacMillan Bloedel Bows to Pressure from Greenpeace," *Financial Times*, 19

June 1998, and from "Natives, Enviros, MacMillan Bloedel Sign Clayoquot Truce," *Environment News Service*, <ens.lycos.com/ens/jun99/1999L-06-17-03.html>, viewed 18 June 1999.

8. Scott Kilman and Helene Cooper, "Monsanto Falls Flat Trying to Sell Europe on Bioengineered Food," *Wall Street Journal*, 11 May 1999; Scott Kilman, "Once Quick Converts, Farmers Begin to Lose Faith in Biotech Crops," *Wall Street Journal*, 19 November 1999; "Monsanto Chief Admits Public Relations Disaster," *Financial Times*, 7 October 1999; Robet B. Shapiro, Monsanto Chairman, Address to Greenpeace Business Conference, London, U.K., 6 October 1999, available at <www.monsanto.com/monsanto/mediacenter/speeches/99oct6_Shapiroscript.html>, viewed 13 December 1999.

9. Control Risks Group, op. cit. note 7; Control Risks Group, <www.crg.com/ControlRisks/crg/index.html>, viewed 13 October 1997.

10. Werner Levi, *Contemporary International Law: A Concise Introduction* (Boulder, CO: Westview Press, 1991); David A. Wirth, "A Matchmaker's Challenge: Marrying International Law and American Environmental Law," *Virginia Journal of International Law*, winter 1992.

11. Peter J. Spiro, "New Global Communities: Nongovernmental Organizations in International Decisionmaking Institutions," *Washington Quarterly*, winter 1995; "Civil Society, The U.N. and the World Bank," *Civicus World*, November/December 1998; P.J. Simmons, "Learning to Live with NGOs," *Foreign Policy*, fall 1998.

12. Wirth, op. cit. note 10; David A. Wirth, "Reexamining Decision-Making Processes in International Environmental Law," *Iowa Law Review*, May 1994; Daniel J. Shepard, "UN Seeks Experts' Testimony in Series of Extraordinary Hearings on Development," *Earth Times*, 15 June 1994.

13. "Session Disrupted, Trade Ministers Insist They Will Continue," *Associated Press*, 1 December 1999; for a skeptical view of NGOs, see Martin Wolf, "Uncivil Society," *Financial Times*, 1 September 1999; on the various roles of NGOs, see Simmons, op. cit. note 11, and Steve Charnovitz, "Two Centuries of Participation: NGOs and International Goverance," *Michigan Journal of*

International Law, winter 1997.

14. Don Hinrichsen, "The Earth Summit," *The Amicus Journal*, winter 1992.

15. United Nations, *Agenda 21: The United Nations Program of Action from Rio* (New York: 1992); Kathryn G. Sessions, "Options for NGO Participation in the Commission on Sustainable Development," UNA-USA Background Paper (Washington, DC: United Nations Association of the United States, May 1993); "About the Steering Committee," NGO Steering Committee to the U.N. C.S.D., <www.csdngo.org/csdngo/steer/sc_index.htm>, viewed 12 December 1999.

16. *Eco* is produced regularly by NGOs at major international negotiations. The *Earth Negotiations Bulletin* is published by the International Institute for Sustainable Development (IISD) of Winnipeg, MN, Canada. It is available at <www.iisd.ca/linkages>.

17. Global Climate Coalition, "Signing Kyoto Protocol a Major Mistake Which Weakens U.S. Position, GCC Says," press release (Washington, DC: 11 November 1998); Sustainable Energy Industries Council of Australia, Inc., brochure; European Business Council for a Sustainable Energy Future, brochure; The Business Council for Sustainable Energy, Washington, DC, brochure; UNEP Insurance Industry Initiative, "Position Statement, Presented at the Conference of the Parties of the United Nations Framework Convention on Climate Change," Kyoto, December 1997.

18. "An Introduction to GLOBE," Global Legislators Organization for a Balanced Environment, <www.globeint.org>, viewed 11 December 1999.

19. Thalif Deen, "Politics: NGOs Barred From Speaking at U.N. Population Meeting," *Inter Press Service*, 1 July 1999; "NGOs and the United Nations," Comments for the Report of the Secretary General, Global Policy Forum, June 1999, available at <www.globalpolicy.org/ngos/docs99/gpfrep.htm>.

20. Commission on Global Governance, op. cit. note 2; Richard Falk and Andrew Strauss, "Globalization Needs a Dose of Democracy," *International Herald Tribune*, 5 October 1999; Erskine Childers with Brian Urquhart, *Renewing the United Nations System* (Uppsala, Sweden: Dag Hammerskjold Foundation, 1994);

"Objectives of the Millennium Forum," <www.millenniumforum.org/html/Mfobject.html>, viewed 12 December 1999.

21. "Clinton's Plea: 'Open the Meetings'," *New York Times*, 1 December 1999; Kahn and Sanger, op. cit. note 2; Steven Greenhouse, "Trade Ministers Sidestep Issue of Secrecy," *New York Times*, 4 December 1999.

22. World Bank, "The World Bank Policy on Disclosure of Information" (Washington, DC: March 1994); Bank Information Center, "Fulfilling the IDA-12 Mandate: Recommendations For Expanding Public Access to Information at the World Bank" (Washington, DC: July 1999); Lori Udall, *The World Bank Inspection Panel: A Three Year Review* (Washington, DC: Bank Information Center, October 1997); Center for International Environmental Law, "World Bank Inspection Panel," <www.igc.apc.org/ciel/wbip.html>, viewed 12 December 1999; on the International Finance Corporation's independent ombudsman, see World Bank, *Environment Matters: Annual Review* (Washington, DC: 1999).

23. Jeffrey Sachs, "IMF is a Power unto Itself," *Financial Times*, 11 December 1997; Nancy C. Alexander, *Finance for Development: A Dialogue with the Bretton Woods Institutions* (New York: Friedrich Ebert Stiftung, 1999); Friends of the Earth–US, *Arming NGOs With Knowledge: A Guide to the International Monetary Fund* (Washington, DC: undated); Jan Aart Scholte, Institute of Social Studies, The Hague, "Civil Society and the International Monetary Fund: An Underdeveloped Dialogue," Paper for Discussions in Washington, DC, May 1998.

24. United Nations, "Secretary-General Proposes Global Compact on Human Rights, Labour, Environment, In address to World Economic Forum in Davos," press release (New York: 1 February 1999).

25. Global Environmental Management Initiative, *Fostering Environmental Prosperity: Multinationals in Developing Countries* (Washington, DC: February 1999); U.N. Conference on Trade and Development, *Self Regulation of Environmental Management: An Analysis of Guidelines Set by International Industry Associations for Their Member Firms* (New York: United Nations, 1996); Control Risks Group, op. cit. note 7; "Lawyers See Rise in Court Cases

Attempting to Apply U.S. Laws Outside Borders," *International Environment Reporter*, 25 June 1997.

26. Figure of 10,000 companies from "ISO 14001 Certificates Predicted to Reach 30,000 Next Year," *ENDS Report*, June 1999; Global Environment & Technology Foundation, "The ISO 14000 Information Guide" (Annandale, VA, undated); IISD, *Global Green Standards: ISO 14000 and Sustainable Development* (Winnipeg, MN, Canada: 1996); Riva Krut and Harris Gleckman, *ISO 14001: A Missed Opportunity for Sustainable Global Industrial Development* (London: Earthscan, 1998).

27. IFOAM from Madden and Chaplowe, op. cit. note 5; Forest Stewardship Council, "Who We Are," information sheet, <www.fscoax.org>, viewed 29 June 1999; Marine Stewardship Council, "Our Empty Seas: A Global Problem, A Global Solution," information brochure (London: April 1999).

28. Wolfgang H. Reinicke, "Global Public Policy," *Foreign Affairs*, November/December 1997; see also <www.globalpublic policy.net>.

INDEX

ABOUT THE AUTHOR

HILARY FRENCH is Vice President for Research at the Worldwatch Institute, a nonprofit policy research organization in Washington, DC, devoted to the analysis of global environment and development issues. Her research and writing focuses on the role of international institutions in environmental protection and sustainable development and on the integration of environmental concerns into international economic policymaking.

During her 12 years at the Institute, Ms. French has written six Worldwatch Papers, including *Investing in the Future: Harnessing Private Capital Flows for Environmentally Sustainable Development, Partnership for the Planet: An Environmental Agenda for the United Nations,* and *Costly Tradeoffs: Reconciling Trade and the Environment.* She has also been a coauthor of nine of the Institute's annual *State of the World* reports, and Associate Project Director for four editions. She contributes to the Institute's other annual publication, *Vital Signs,* and is a member of the Editorial Board of *World Watch,* the Institute's bimonthly magazine, where her articles appear regularly. She has written articles for numerous other publications, and lectured widely at conferences and universities in the United States and abroad.

Prior to joining the Institute, Ms. French interned with the U.N. Development Programme in Côte d'Ivoire and with the United Nations Institute for Disarmament Research in Geneva. She holds degrees from Dartmouth College and from the Fletcher School of Law and Diplomacy.